CAMBRIDGE COUNTY GEOGRAPHIES

General Editor: F. H. H. GUILLEMARD, M.A., M.D.

GLOUCESTERSHIRE

Cambridge County Geographies

GLOUCESTERSHIRE

by

HERBERT A. EVANS, M.A.

With Maps, Diagrams and Illustrations

Cambridge :
at the University Press
1925

CAMBRIDGE UNIVERSITY PRESS
Cambridge, New York, Melbourne, Madrid, Cape Town,
Singapore, São Paulo, Delhi, Mexico City

Cambridge University Press
The Edinburgh Building, Cambridge CB2 8RU, UK

Published in the United States of America by Cambridge University Press, New York

www.cambridge.org
Information on this title: www.cambridge.org/9781107697393

First edition 1909
Second edition 1914
Reprinted 1925
First published 1925
First paperback edition 2013

A catalogue record for this publication is available from the British Library

ISBN 978-1-107-69739-3 Paperback

CONTENTS

ILLUSTRATIONS

MAPS

The Author desires to thank his friend, Mr James G. Wood, M.A., F.G.S., for the section treating of the geology of the County.

H. A. E.

1. County and Shire. Origin and Meaning of the Words.

If we take a map of England and contrast it with a map of the United States, perhaps one of the first things we shall notice is the dissimilarity of the arbitrary divisions of land of which the countries are composed. In America the rigidly straight boundaries and rectangular shape of the majority of the States strike the eye at once ; in England our wonder is rather how the boundaries have come to be so tortuous and complicated—to such a degree, indeed, that until recently many counties had outlying islands, as it were, within their neighbours' territory. We may guess at once that the conditions under which the divisions arose cannot have been the same, and that while in America these formal square blocks of land, like vast allotment gardens, were probably the creation of a central authority, and portioned off much about the same time ; the divisions we find in England own no such simple origin. Our guess would not have been wrong, for such, in fact, is more or less the case. The formation of the English counties in many instances was (and is—for they have altered up to to-day) an affair of slow growth. King Alfred is credited with having made them, but inaccurately, for some existed before his time, others not till long after his death, and their origin was—as their names tell us—of very diverse nature.

Let us turn once more to our map of England. Collectively, we call all our divisions counties, but not every one of them is accurately thus described. Gloucestershire, as we shall see, is not. Some have names complete in themselves, such as Kent and Sussex, and we find these to be old English kingdoms with but little alteration either in their boundaries or their names. To others the terminal *shire* is appended, which tells us that they were *shorn* from a larger domain—*shares* of Mercia or Northumbria or some other of the great English kingdoms. The term county is of Norman introduction,— the domain of a *Comte* or Count.

County and shire are now practically synonymous, and we can talk of "Gloucestershire" or "the county of Gloucester," but not of "the county of Gloucestershire."

Gloucestershire is a shire of the ancient great central Anglian kingdom of our land—Mercia. The chief towns of Mercia had long been military centres, their governors levying men from the surrounding district, and in 918, when Ethelfleda (Æthelflead) was Lady of the Mercians, we first hear of Gloucester as one of these centres. The name "Gloucestershire" is not heard of till 1016, when the Saxon Chronicle records that after the battle with the English at Ethandun "went King Cnut up with his army to Gloucestershire where he learned that King Eadward was." Now the fact that we have no mention of the shire for nearly 100 years after the first mention of the town makes it probable that Mercia was first divided into shires at the time of the last Danish invasion, about the year 1000, when measures were being taken to resist the invaders, and the "hundreds"—about which we shall learn later—were grouped into shires, each group taking its name from its chief town. Thus the division into shires was primarily a military measure,

Gloucester Cathedral

but, as we shall see, the boundary of a particular shire was liable to be modified to suit the convenience of great landowners, especially if they were great monastic houses.

As to the name of the town which gives its name to the county—the Welsh Gloui (Caergloui) became the Roman Glevum, and this the Saxon Gleauanceaster, now Gloucester.

2. General Characteristics. Position and Natural Conditions.

What we now call the Bristol Channel was in pre-historic times a marsh through which the Severn found its way to the Atlantic ; but when the land subsided the sea took its place and drowned the ancient bed of the river, till at high water it pushed up to Gloucester and even beyond. At low water you may see the shoals and rocks which mark the present course of the river when unaffected by the tide.

Now the fact that it is penetrated by an arm of the sea gives Gloucestershire a unique position among English counties. If we look at other counties which have a tidal river, Kent, Essex, Lincolnshire, Lancashire, we find that they all have a sea coast and are in the fullest sense maritime counties, but Gloucestershire has no coast at all. It is essentially a midland county, yet at the same time its tidal river gives it many of the advantages of a maritime county, for large merchantmen are carried by the flowing tide to Bristol and—with the aid of the Berkeley canal—to Gloucester.

But, like other midland counties, its main features are agricultural. The hilly districts grow corn, and pasture sheep, while the lowlands consist of a rich deep soil which is divided between orchards and large dairy farms. This description, however, applies to the country east of the Severn only. West of the river we find ourselves in another world : dairy farms and apple orchards still abound, but the character of the scenery and the accent of the natives soon tell us that we have reached the borderland of Wales. Moreover, about half of this peninsula between the Wye and Severn is covered by forest, and here are coal and iron mines, which have given rise to the industrial settlements of Coleford, Lydney, and Cinderford.

Thus we see that, alone among midland counties, Gloucestershire is at once agricultural, industrial, and

Golden Valley

maritime. If a man were to travel in a straight line from Burford, on the Oxfordshire border of the county, to Monmouth, or from Moreton to Chepstow, the character of the country he would pass through would be more diversified than in any other journey of the same length in England.

3. Size. Shape. Boundaries. Detached Portions.

The length of the county from Clifford Chambers to Bitton is 60 miles, its breadth from Lechlade to Lancaut, not including the estuary of the Severn which is three miles across, is 40 miles. The circumference of the county, including the two sides of the estuary up to Awre, is about 270 miles, and this encloses an area of 1258 square miles or 805,482 acres.

The shape of Gloucestershire is extremely irregular. On the north and north-west its outline is very much indented, but we may describe it roughly as a five-sided figure, of which the south side is the shortest.   Beginning from the mouth of the Bristol Avon and proceeding northwards we may define its sides as follows: (1) from the mouth of the Avon to Preston, near Ledbury, (2) from Preston to Clifford Chambers, (3) from Clifford Chambers to Lechlade, (4) from Lechlade to the Shire Stones near Marshfield, and (5) from Marshfield to the mouth of the Avon.

The boundaries of the county correspond with those of the southern portion of the ancient diocese of Worcester, with the addition of the Forest Peninsula, which was in the diocese of Hereford. The adjacent counties, beginning at the mouth of the Wye, are Monmouth, Hereford, Worcester, Warwick, Oxford, Berks (for about two miles near Lechlade), Wilts, and Somerset. With the exception of the Avon on the south, and the Wye on the west, the county boundary is rarely formed by rivers; the northern boundary, which separates the county from Worcestershire and divided the old diocese of Worcester into two parts, is extremely irregular. This irregularity

The Shire Stones near Bath, where the counties of Somerset, Wilts. and Gloucester meet

is chiefly due to the fact that the ancient estates of the church of Worcester were included in Worcestershire, and those of the church of Deerhurst—Todenham and Little Compton—in Gloucestershire.

There were formerly five detached portions of the

county, but about 60 years ago they were incorporated with the counties in which they are situated, but not always with the diocese; thus, Little Compton and Sutton-under-Brailes, now in Warwickshire, remain in the diocese of Gloucester; and Sherrington, now in Oxfordshire, was long in the diocese of Worcester. The other two portions are Widford, now in Oxfordshire; and the greater part of Minety, now in Wiltshire. For the transference the consent of both the counties interested was necessary and this was not always obtained. Hence we have in Gloucestershire the long Worcestershire island of Cutsdean, and except for a mile on the north-east, where it is bounded by Warwickshire, another Worcestershire island of Blockley. Indeed so mixed are the counties near Moreton that a well-known monument, the four shire stone, marks the meeting of the four counties of Worcester, Warwick, Oxford, and Gloucester.

Like the outliers of the Cotswolds which occur in the vales of Gloucester and Evesham, so these islands were outlying portions of the possessions of some landowner of the adjacent county, and they would also belong to the "hundred" in which this landowner's chief possessions were situated. Thus Little Compton, belonging to the priory of Deerhurst, was in Deerhurst hundred, and Shennington, belonging to the abbey of Tewkesbury, in that of Tewkesbury. The island of Cutsdean is a chapelry of the parish of Bredon in Worcestershire, at least twelve miles distant, and was given by Offa in the eighth century to the monastery which his grandfather had founded there. It afterwards belonged to the priory

of Worcester, but has always remained parochially attached
to Bredon.   It is therefore in the same "hundred," but
of the hundreds we shall have to write in a later chapter.

4. Surface and General Features.

On a general view the county falls into three main
divisions—hill, vale, and forest—each division running

Cleeve Hill, near Cheltenham

from north-east to south-west.   It must not be supposed,
however, that there are no vales in the hill district or no
hills in the forest.   The Hill district, then, extends from
Chipping Campden to Bath ; it is a rough stony country,
formerly consisting almost wholly of sheep-down, but now
for the most part brought under the plough.   Its general

altitude varies from 900 to 550 feet, but at Cleeve Cloud, near Cheltenham, it reaches a height of 1070 feet, and at Broadway Hill on the old " Buckle Street," a height of 1048 feet.   From these points extensive views may be

Cross at Condicote

obtained across the Vale to the Malverns and the Welsh mountains, while behind the spectator lies a vast tract of upland divided into enclosures by stone walls, and penetrated by soft green valleys watered by clear and gentle streams.   Here and there tufts of firs have been planted, while the sides of the valleys are often clothed with hanging woods. It is in these valleys that the villages are mainly situated, but a few like Condicote and Cutsdean, mentioned in the last chapter, lie up on the open down. These hills have always been a favourite field for the chase of wild animals. They are now hunted by several packs of foxhounds, but in early times, in Shakespeare's day for example, they were the home of the wild deer. But now, as then, you may walk for miles without seeing a house until you drop abruptly upon some village or out-lying farmstead. Southwards as far as the Stroudwater this hill district is known as the Cotswolds, but the part south of this river, though the same name is commonly given to the whole range, is sometimes called the Stroud-water Hills. To the west these hills form a steep escarp-ment which rises abruptly from the vale, and when looked at from the other side of the Severn they form a long unbroken line bounding the horizon : to the east and south-east they sink gradually into the basins of the Thames and Bristol Avon.

Our next division, the Vale, is the whole district lying between the Severn and the hills, from Evesham and Tewkesbury to the Bristol Avon. It is sometimes divided into the vales of Evesham, Gloucester, and Berkeley. The first is really the valley of the Warwickshire Avon, and most of it is included in Worcestershire and Warwick-shire; the second extends from Tewkesbury to Gloucester; and the third comprises the rest of the vale, but these divisions are rather imaginary and convenient than actual. The soil is deep and rich, and famous for its produce.

View of the Cotswolds from above Newnham

Dairy farms and orchards, interspersed with arable fields, cover most of the surface; in the former is made the famous Gloucestershire cheese, and in the latter grow the apples which make the no less famous Gloucestershire cider.   The cheese made in the vale of Berkeley, where the pastures are particularly rich, is the kind called double Gloucester.   In the lower parts of this vale, south of Berkeley, between the hills and the river, is a tract of undulating country, from 200 to 300 feet in altitude, hardly less fertile than the vale itself, and varied with woods and plantations.

The Forest is the peninsula between the Severn and the Wye.   It is many centuries since the whole district was covered by trees.   The part which may now practically (though not technically) be termed the forest reaches from Mitcheldean to Lydney, a distance of 10 miles, and a straight line carried through the centre from Newnham to the Wye, near Monmouth, would also be 10 miles. A long backbone of hills runs down the peninsula from May Hill, near Longhope, to Tidenham, which is the principal feature of the landscape from the other side of the Severn.   These hills vary from 400 to 900 feet in height.   May Hill itself is 969 feet, and the highest point of Tidenham Chace reaches 778 feet.   Beyond the Wye, which has worked its way through it, this range extends in a south-westerly direction as far as Caerleon in Monmouthshire.   The whole chain from May Hill to Caerleon is 33 miles in length, and for the beauty and variety of its surface, the richness of its fauna and flora, and the wide sweep of land and water commanded by its numerous elevations, it is one of the most delightful parts of all England.

From this Forest of Dean, as it is called, was formerly drawn a considerable part of the timber for the navy, but by 1783 the greater part of it had been felled and the site had become a barren heath. Since then these waste

The Speech House, Forest of Dean

places have been replanted, and the forest at the present day mainly consists of oaks about 100 years old. Here and there, however, there are clumps of aged beeches, and near the Speech House is a large thicket of some of the oldest hollies to be found in England. Here too are many veteran oaks, and others are sprinkled over the whole forest. This Speech House is situated in the very centre of the forest and is the place of meeting of the Court Verderers, who manage the "Vert and Venison" of the forest under the Crown. The deer were all killed off half a century ago, so that the Vert, i.e. the timber and pasture, alone remains. This court has been held ever since the introduction of the Norman forest law. Of the mines we shall have to speak in a later chapter; they have of course left their mark upon the appearance of the forest where they occur, and round Cinderford and Coleford one might almost be in the mining districts of Cheshire or South Wales. In spite of this the general aspect is that of lofty wooded hills separated by deep dales, down which run streams generally reddened by iron ore. Here and there are open commons on which the cottages of the miners are built.

There remains the spacious projection of the county north of May Hill watered by the river Leadon, which deserves to rank as a fourth division. It is rich in meadows, orchards, and woodlands, and in its general features resembles the adjacent parts of Herefordshire. It might almost be called a little county of itself, with Newent for its capital.

5. Watersheds and Rivers.

The principal watershed is the range of the Cotswolds, which separates the basins of the Severn and the Thames. Owing to the slope of these hills to the south-east most of the streams north of Stroud flow into the Thames—the

Thames itself, the Churn, the Coln, the Leach, and the Windrush. The Thames, which has no more claim than the Bristol Avon or the Wye to be regarded as a

Thames Head in Winter

Gloucestershire river, is generally considered to have its source at Thames Head, near Kemble; other people claim the Seven Springs at Coberley, near Cheltenham, or Ullen farm as the source, as the farthest point from the mouth;

but, call it Thames or Churn, a copious outflow issues from these springs and, after watering one of the most picturesque valleys in the Cotswolds, flows round Cirencester, and at Cricklade in Wiltshire is unmistakably the Thames.

The Coln rises at Charlton Abbots, flows past Andoversford, Withington, Bibury, and Fairford, and

Seven Springs, near Cheltenham

joins the Thames at Lechlade. The famous Roman station of Chedworth is upon its banks.

The Leach rises near Northleach, and after flowing between the twin villages of Eastleach (Turville and Martin) joins the Thames just below Lechlade. Some two miles below the confluence, the Thames, which for a short distance here divides Gloucestershire from Berkshire, takes its final leave of our county.

The Windrush, the longest of all these rivers, rises two miles above Temple Guiting, leaves the county at Great Barrington, and passing Burford and Witney flows into the Thames a few miles below Shifford.

The Windrush above Bourton

The Warwickshire Avon, to come to the rivers flowing into the Severn, has not much to do with Gloucestershire. For five miles below Stratford-on-Avon, and again for four miles above Tewkesbury—where it joins the Severn—it forms the boundary of the county. Near Stratford it is joined by the Stour (the commonest of our river names and probably, like Avon, meaning "river" or "water"), which for a short distance in the latter part of its course claims to be a Gloucestershire river. A couple of miles above Shipston it receives the Knee Brook, a rivulet which rises in the hills above Chipping Campden, and for about three miles divides the county from Warwickshire and Worcestershire. A glance at the map will show how complicated the county

The Meeting of the Severn and Avon at Tewkesbury

boundaries are at this point. But the chief Gloucestershire tributary of the Avon is the Isbourne: this rises at Charlton Abbots, only half a mile from the source of the Coln, and flows northwards past Winchcombe and Toddington to join the Avon at Evesham. The little ridge which here separates the basins of the Thames and Severn forms a capital study of a watershed in miniature.

The Severn itself enters the county at Tewkesbury; and two miles above Gloucester, not far from the point to which the tide reaches, it separates into two branches which unite again opposite Gloucester and form the fertile Isle of Alney. It then follows a very sinuous and gradually widening channel as far as Framilode, where it suddenly bends westwards and flows past Newnham round to Fretherne, a course of ten miles, thus forming a peninsula the neck of which is only a mile and a quarter across. Below Fretherne the river expands considerably, to a width of a mile and a half or more, and then with two contractions, at Sharpness and at Beachley, it loses itself in the Bristol Channel. The exact place where the river ends and the channel begins is merely a question of names, but it is convenient to speak of "the Severn" above the mouth of the Avon, and of the "Bristol Channel" below it.

There is a remarkable phenomenon in connexion with the tides on the Severn which though seen on other rivers is here unusually well-marked. This is the "bore." A man standing in Awre churchyard would at low water look down upon a large sandbank, called the Noose, filling up the greater part of the bed of the river, and leaving only a narrow channel for the water on either side of it. At spring tides the body of water rushing in from the Atlantic acquires an increased velocity as it is contracted in its passage through the narrow strait at Sharpness; then as soon as it strikes this Noose island it is backed and ponded up. But it soon rises high enough to cover the sandbank, and then rushing on with greater violence than ever it encounters the sudden contraction of the river bed

at Fretherne.   This causes it to mount up to the height
of three feet or more, and to rush roaring up the narrowed
channel.   Frank Buckland's vivid account of the phenomenon is worth quoting:—"In a few minutes I saw a
curved white line stretched right across the channel
coming round the corner of the river.   With a fearful
velocity this white line advanced steadily up the river, and
as it neared us I saw that it consisted of a wave about
three feet high curling over with foam at its summit, and
forming a distinct wall....The roar that it made was like
nothing I ever before heard....Behind the first wave wall
came a second, and then a third, and then the full body
of the tide boiled like a caldron.   Behind this again swept
a broad sheet of water—the main army of the flood in
rear of the advance guard."   Smaller bores occur at every
Spring tide, but the highest are at the "Palm tides" in
March.

After the Avon the Severn receives on its left bank
the Chelt, the Frome or Stroudwater, and the Little
Avon.   The Chelt rises at Dowdeswell not far from
Andoversford, and flowing westward gives its name to
Cheltenham and falls into the Severn between Deerhurst
and Ashleworth.

The Frome rises in the hills above Miserden, makes a
bend to the west at Saperton, passes Stroud and Stonehouse, and reaches the Severn at Framilode; in its
comparatively short course it descends some 750 feet.
The convenient situation of the Stroud valley and the
suitable nature of the river water long ago made this the
chief centre of the clothing trade in the western midlands.

The Little Avon rises at Newington Bagpath, issues into the level country at Alderley and flows in a northerly direction to join the Severn at Berkeley. The little creeks into which the tiny brooks flow on both sides of the estuary are called " pills."

As for the Avon, it has but the faintest claim to be reckoned among the Gloucestershire rivers : it rises at Tetbury, forms the county boundary for a couple of miles, and then starts away on a long tour through Wilts and Somerset to touch our county again at Bitton and bound it as far as to the Bristol Channel. On its way it takes up on its right bank three little rivers, the Boyd, the Frome, and the Trym.

The only Gloucestershire tributary of the Severn on its right bank is the Leadon. This rises at Evesbatch in Herefordshire at a height of 551 feet, skirts that county a couple of miles below Ledbury, passes Dymock, forms the Worcester boundary for three or four miles, and finally enters the Severn at Alney Island.

The Wye, a Welsh river, forms the boundary of the county for about four miles near English Bicknor, and again from Redbrook to Beachley. For most of the way it is accompanied on the English side by Offa's Dyke, and the precipitous cliffs, along the summit of which the dyke is carried, seem to warn the Welshman to advance no farther,—in fact the English county seems in scorn to turn its back upon the Celt.

The Peninsula of Lancaut, with the Junction of the Wye and Severn in the Distance

6. Geology and Soil.

By Geology we mean the study of the rocks, and we must at the outset explain that the term *rock* is used by the geologist without any reference to the hardness or compactness of the material to which the name is applied; thus he speaks of loose sand as a rock equally with a hard substance like granite.

Rocks are of two kinds, (1) those laid down mostly under water, (2) those due to the action of fire.

The first kind may be compared to sheets of paper one over the other. These sheets are called *beds*, and such beds are usually formed of sand (often containing pebbles), mud or clay, and limestone, or mixtures of these materials. They are laid down as flat or nearly flat sheets, but may afterwards be tilted as the result of movement of the earth's crust, just as you may tilt sheets of paper, folding them into arches and troughs, by pressing them at either end. Again, we may find the tops of the folds so produced wasted away as the result of the wearing action of rivers, glaciers, and sea-waves upon them, as you might cut off the tops of the folds of the paper with a pair of shears. This has happened with the ancient beds forming parts of the earth's crust, and we therefore often find them tilted, with the upper parts removed.

The other kinds of rocks are known as igneous rocks, which have been melted under the action of heat and become solid on cooling. When in the molten state they have been poured out at the surface as the lava of volcanoes, or have been forced into other rocks and cooled

in the cracks and other places of weakness. Much
material is also thrown out of volcanoes as volcanic ash
and dust, and is piled up on the sides of the volcano.
Such ashy material may be arranged in beds, so that it
partakes to some extent of the qualities of the two great
rock groups.

The production of beds is of great importance to
geologists, for by means of these beds we can classify the
rocks according to age. If we take two sheets of paper,
and lay one on the top of the other on a table, the upper
one has been laid down after the other. Similarly with
two beds, the upper is also the newer, and the newer will
remain on the top after earth-movements, save in very
exceptional cases which need not be regarded by us here,
and for general purposes we may regard any bed or set of
beds resting on any other in our own country as being
the newer bed or set.

The movements which affect beds may occur at
different times. One set of beds may be laid down flat,
then thrown into folds by movement, the tops of the
beds worn off, and another set of beds laid down upon the
worn surface of the older beds, the edges of which will
abut against the oldest of the new set of flatly deposited
beds, which latter may in turn undergo disturbance and
renewal of their upper portions.

Again, after the formation of the beds many changes
may occur in them. They may become hardened, pebble-
beds being changed into conglomerates, sands into sandstones, muds and clays into mudstones and shales, soft
deposits of lime into limestone, and loose volcanic ashes

	Names of Systems	Subdivisions	Characters of Rocks
TERTIARY	Recent / Pleistocene	Metal Age Deposits / Neolithic ,, / Palaeolithic ,, / Glacial ,,	Superficial Deposits
	Pliocene	Cromer Series / Weybourne Crag / Chillesford and Norwich Crags / Red and Walton Crags / Coralline Crag	Sands chiefly
	Miocene	Absent from Britain	
	Eocene	Fluviomarine Beds of Hampshire / Bagshot Beds / London Clay / Oldhaven Beds, Woolwich and Reading / Thanet Sands [Groups]	Clays and Sands chiefly
SECONDARY	Cretaceous	Chalk / Upper Greensand and Gault / Lower Greensand / Weald Clay / Hastings Sands	Chalk at top / Sandstones, Mud and Clays below
	Jurassic	Purbeck Beds / Portland Beds / Kimmeridge Clay / Corallian Beds / Oxford Clay and Kellaways Rock / Cornbrash / Forest Marble / Great Oolite with Stonesfield Slate / Inferior Oolite / Lias—Upper, Middle, and Lower	Shales, Sandstones and Oolitic Limestones
	Triassic	Rhaetic / Keuper Marls / Keuper Sandstone / Upper Bunter Sandstone / Bunter Pebble Beds / Lower Bunter Sandstone	Red Sandstones and Marls, Gypsum and Salt
PRIMARY	Permian	Magnesian Limestone and Sandstone / Marl Slate / Lower Permian Sandstone	Red Sandstones and Magnesian Limestone
	Carboniferous	Coal Measures / Millstone Grit / Mountain Limestone / Basal Carboniferous Rocks	Sandstones, Shales and Coals at top / Sandstones in middle / Limestone and Shales below
	Devonian	Upper / Mid / Lower } Devonian and Old Red Sandstone	Red Sandstones, Shales, Slates and Limestones
	Silurian	Ludlow Beds / Wenlock Beds / Llandovery Beds	Sandstones, Shales and Thin Limestones
	Ordovician	Caradoc Beds / Llandeilo Beds / Arenig Beds	Shales, Slates, Sandstones and Thin Limestones
	Cambrian	Tremadoc Slates / Lingula Flags / Menevian Beds / Harlech Grits and Llanberis Slates	Slates and Sandstones
	Pre-Cambrian	No definite classification yet made	Sandstones, Slates and Volcanic Rocks

into exceedingly hard rocks. They may also become cracked, and the cracks are often very regular, running in two directions at right angles one to the other. Such cracks are known as *joints*, and the joints are very important in affecting the physical geography of a district. Then, as the result of great pressure applied sideways, the rocks may be so changed that they can be split into thin slabs, which usually, though not necessarily, split along planes standing at high angles to the horizontal. Rocks affected in this way are known as *slates*.

If we could flatten out all the beds of England, and arrange them one over the other and bore a shaft through them, we should see them on the sides of the shaft, the newest appearing at the top and the oldest at the bottom, as in the annexed table. Such a shaft would have a depth of between 10,000 and 20,000 feet. The strata beds are divided into three great groups called Primary or Palaeozoic, Secondary or Mesozoic, and Tertiary or Cainozoic, and the lowest Primary rocks are the oldest rocks of Britain, which form as it were the foundation stones on which the other rocks rest. These may be spoken of as the Pre-Cambrian rocks. The three great groups are divided into minor divisions known as systems. The names of these systems are arranged in order in the table. On the right hand side, the general characters of the rocks of each system are stated.

With these preliminary remarks we may now proceed to a brief account of the geology of the county.

Gloucestershire contains a longer range of stratified rocks than any other county; and so presents a typical field for the study of geology. The igneous or unstratified rocks however occur only in a limited area near Tortworth and Charfield in the south-west part of the county where the Greenstone has been forced up through the stratified Llandovery rocks of the Upper Silurian series.

The rocks in this county follow a law which will be found to prevail generally throughout England and Wales; that is that the older rocks, which were first deposited and so are lowest down among the series, are to be found to the westward; and "dip" or slope downwards towards the east and (except in cases where they cease altogether) pass under and are overlaid by the newer rocks. As we pass eastward across the county we are therefore continuously passing over newer and newer formations.

This regularity however is broken in one line by a great faulting or dislocation connected with the formation of the Malvern Hills. This line passes nearly due N. and S. commencing at the Abberley Hills in Worcestershire, follows the line of the Malverns, enters our county near Ledbury, continues through May Hill, crosses the Severn near Purton Passage, and passing by Tortworth reaches Mangotsfield.

This dislocation has had a most important effect. It has brought to the surface, near the point where it enters the county and again near Huntley on the flanks of May Hill, rocks of Pre-Cambrian age; that is of the oldest series known in England or Wales; and also the upper Silurian Beds in the May Hill and Berkeley districts; and has let up the intrusive Greenstone dykes at Tortworth already mentioned.

With these exceptions the formations in the county proceed regularly in an ascending scale from the Old Red Sandstone on the banks of the Wye and Avon in the west, to the Oxford Clay—the newest formation to be found in the county—on the banks of the Coln and Thames near Fairford in the east.

Another geological fact which has produced prominent physical features in the aspect of the county is that by reason of elevations in one direction and depressions in the other, due to upheavals and shrinkages, the edges of the lower beds or strata are commonly found to be tilted up to higher actual elevations than any part of the more recent beds that in other parts overlie them. Thus the Old Red Sandstone at the Buckstone on the west of Dean Forest attains a height exceeded west of Severn only by the still older Silurian rocks of May Hill. Again on the east of Severn the summits of the highest points of the Cotswold range, or the hills that originally belonged to it, such as Bredon, Cleeve Cloud, Leckhampton, Birdlip and Kimsbury, are composed of the lower beds of the Oolitic series; while the upper and newer beds of the same series which lie on the slopes of the Cotswolds eastwards never attain the same elevation.

We will now consider the extent and situation of the various deposits as they are exposed on the surface beginning from the west.

The Old Red Sandstone is to be found on the banks of the Wye between Tintern and Redbrook. Northward, southward, and eastward it immediately dips under and is overlaid by the Carboniferous series of the Dean

Forest, at the base of which is the Carboniferous Limestone, forming the magnificent cliffs near Symond's Yat
and between Tintern and Chepstow. This again dips
under the Millstone Grit, which caps Tidenham Chace;
and that again is overlaid within the Forest area by the
coal measures, or rocks which contain the coal beds. In
this case the whole series rise again and crop out along
the eastern face of the Forest hills over the Old Red and
are cut out by the Severn valley. The limestone however
continues southward to Tidenham and then turns eastward across the Severn and takes in and passes under the
Bristol coalfield, as it did the Forest coalfield, and again
returns by the Avon cliffs, crosses the Bristol Channel
into Monmouthshire, and so reunites with the cliffs on the
Wye; but its passage under the Severn in both cases cannot be seen.

These older series do not recur again in the county.
In places the Old Red (denuded of its burden of the coal
measures) passes under, and in other places is cut off by
a fault against, the New Red Marls and Sandstones which
are exposed in the cliffs of Aust, Sedbury, and Westbury-
on-Severn (capped however in all three cases by Lias clays
and limestone) and cover a large tract of country in the
valley of the Leadon.

We next find a broad band of Lias Clay extending
along the western base of the Cotswold range in an
unbroken line from near Tewkesbury to the county
boundary near Bath. That this extended originally much
further westward is apparent from its being found in
many isolated positions towards, and along, the Severn.

The Devil's Chimney, Cheltenham in the distance

The Lias Clay then passes under the Midford Sands, a valuable water-bearing stratum which throws out its water over the underlying clay and so gives rise to a number of springs and streams along the face of the hills.

We have now reached the base of the Oolitic or free-stone series; the lowest bed of which (called therefore the Lower Oolite) forms the "Edge" by which name the escarpment of the Cotswolds has been known from the earliest times[1]. The face of the range is deeply furrowed by combes and valleys formed by the washing out of the underlying clays and sands, which are consequently exposed in them (notably in the Stroud Valley) at great distances from the face of the range.

Passing eastward we meet in succession and over large surfaces the Great Oolite—usually called Bath Stone and largely quarried at Box and Corsham, just beyond the county boundary—and the Forest Marble and the Corn-brash, which extend to the county boundary except so far as they are overlaid by the Oxford Clay near Fairford, as already mentioned.

Considerable parts of this district are, however, so deeply furrowed by the Windrush river and its tributaries that the underlying Midford Sands and Lias Clay are again exposed in the bottoms of their valleys and form the source of many streams which run into the Thames.

This is not the place to speak of fossils; they would

---

[1] Hence Edgehill in Warwickshire, Weston-sub-Edge near Campden, the Edge near Stroud and Wotton-under-Edge, besides other instances.

require a volume to themselves. It must suffice to explain that fossils are animals and plants, either of extinct species or of species still surviving, embedded in the various strata

Westington Quarry, near Campden

of rock. Igneous rocks of course have no fossils. The pit heaps in the Forest afford good specimens of the ferns and other vegetation of the coal period; the cliffs of Aust and Westbury-on-Severn are noted for their remains of fishes and the animals of the Triassic and Liassic periods; and the Oolitic series supply almost endless varieties of the ammonite.

7. Natural History.

Various facts, which can only be shortly mentioned here, go to show that the British Isles have not existed as such, and separated from the continent, for any great length of geological time. Around our coasts, for instance, are in several places remains of forests now sunk beneath the sea, and only to be seen at extreme low water. Between England and the continent the sea is very shallow, but a little west of Ireland we soon come to very deep soundings. Great Britain and Ireland were thus once part of the continent, and are examples of what geologists call recent continental islands. But we also have no less certain proof that at some anterior period they were almost entirely submerged. The fauna and flora thus being destroyed, the land would have to be restocked with animals and plants from the continent when union again took place, the influx of course coming from the east and south. As, however, it was not long before separation occurred, not all the continental species could establish themselves. We should thus expect to find that the parts in the neighbourhood of the continent were richer in species and those furthest off poorer, and this proves to be the case both in plants and animals. While Britain has fewer species than France or Belgium, Ireland has still less than Britain.

From what has already been said about the diversity of the surface of Gloucestershire, we should expect that the fauna and flora would be extensive, and this is just what we find. The wide table-lands of the Cotswolds, the deep combes with which they are penetrated, the rich pastures of the Vale, the tidal waters of the Severn, the heathy commons of the Forest, and the steep rocky woods of the limestone districts on either side of the Severn, have each their characteristic plants, and each have their own attractions to offer to wandering bird or beast. In some of the Cotswold combes, to mention the rarer animals only, the badger still makes his home, and in the dense rocky woods along the Wye he still has an occasional marten and the polecat as his neighbours. The deer in the Forest were destroyed more than half a century ago, on account of the encouragement they gave to poachers— an expedient easy and effectual. In Norman times the Forest was a favourite hunting ground of our kings. On the Cotswolds too, then unenclosed, the red deer was hunted, as we may read in Mr Justice Madden's delightful book, *The Diary of Master William Silence.* In those days too the great bustard, which still roamed over the downs, was hunted with greyhounds. The beasts of the chase are now the fox, the hare, and the otter. Half a dozen packs of foxhounds, of which the Duke of Beaufort's is the most famous, hunt the country east of the Severn. West of the river, though the fox abounds, it is not hunted owing to the natural difficulties of the country. Hares are not so abundant as they used to be, but they have been coursed on the downs ever since the time when

Justice Shallow's fallow greyhound was "outrun on Cotsall."

As for the larger birds of prey Gloucestershire has the same story to tell as other counties, viz. gradual extinction owing to game preserving, and to the decrease of woods and wastes. In the Forest peninsula however some of them survived later than in most parts of England, and sixty or seventy years ago the kite was occasionally to be seen soaring over the open spaces in the Forest and the rough country to the south of it. Ravens nested in the cliffs overhanging the Wye in Tidenham, and were only driven away by the making of the Wye Valley Railway in 1873.

The valley of the Severn being a continuation of the migration route from the Wash south-westwards we are visited by most of the spring and winter migrants; some of the rarer ones, such as the ring ousel, which goes to the hills of Devon or Wales to breed, rest with us for a few days on their journey, wherever they find a supply of their favourite food. All the commoner warblers may be seen in their proper haunts; the nightingale for example, which shuns the bleak uplands of the wolds, may be heard all over the Newent district and in every copse on either side of the forest down to Beachley. This is one of the most westerly districts reached by this bird.

The shores of the Severn are visited by the ordinary shore-birds and waders, gulls, plovers, dunlins, geese and ducks, but the characteristic bird of the grassy mud-flats known locally as *wharfs*, is the sheld-duck, or, as the natives call it, the burrow-duck, from its habit of nesting in a rabbit warren. It arrives in April and stays till the autumn. One of its favourite nesting-places is in the red cliffs at Tidenham, in the lower part of which are deep holes, just above high-water mark, exactly adapted for its home.

There is no very characteristic bird of the Vale that we need mention, for the rarer waders have long ago departed, nor is there any remarkable bird to be noticed on the wolds, now that the stone-curlew has practically disappeared; but the combes watered by clear and sheltered streams are the favourite haunt of the dipper, and here too the beautiful gray wagtail sometimes makes its nest.

Few counties are richer in wild flowers. The road-sides and uncultivated parts of the Cotswolds are clothed with thyme, milkwort, and speedwell, while acres of sainfoin make a blaze of colour unlike anything else in the county. Then on the sides of every disused quarry and every broken bank may be found the handsome woolly-headed thistle (*Cnicus eriophorus*), a plant which unlike most of its fellows soon disappears before cultivation. In the rest of the county it is very seldom found. On some of the grassy slopes several of the rarer orchises occur, and in the darker beechwoods the curious scaly monotropa or "yellow bird's nest." The beech is the characteristic tree of the Cotswolds, and the one that thrives best in its stony soil; the ash also flourishes where it can get a deeper hold. But the great stretches of the surface of the hills would now be quite bare of trees if it were not for the massive clumps and lines of firs which have been planted in modern times.

Along the Leadon, in the Newent district, the meadows in early spring are yellow with daffodils, and from this district southwards to the extremity of the Forest peninsula is the paradise of the Gloucestershire flora. The southern portion of the peninsula alone, south of St Briavels and Lydney, contains over 760 species of flowering plants out

The Double Bend of the Wye as seen from Winter's Leap, Tidenham

of some 2000 species found in the United Kingdom, and 25 kinds of fern. It would probably be difficult to find in England (Wales is another matter) an area of the same size containing so large a number of plants: the whole of Oxfordshire has only about 800. The fact is that, like Herefordshire, its flora has a western and south-western character, witness the abundance of the mistletoe and the pennywort (Cotyledon). Then there is the diversity of the soil—limestone rock, sandstone, rich loam and salt marshy flats, the steep damp woods overhanging the Wye, the open heath at their summit, and the rough copses and meadows which slope down towards the Severn.

We cannot go into any details here respecting the insects of the county, but to take one order alone, the Lepidoptera (butterflies and moths) are very numerous. That this should be so is only what might be anticipated from the diversity of the flora which supplies the food of the caterpillars. More than half the British butterflies may be found without much difficulty, and several other species have been recorded. The most notable of them all is the Large Blue (*Lycaena Arion*) which occurs here and there in the Cotswolds. The number of moths is also considerable, though no doubt less than that of many counties in the east of England where large marshy tracts are left.

8. The Estuary of the Severn.

An estuary, it has been well said, is a drowned river valley. There must therefore have been a time when the Severn below Gloucester was not much wider than it was above, and finally lost itself in a vast morass, which then occupied the space now filled by the Bristol Channel. In a later age a depression of land surface following upon the glacial period let in the waters of the ocean, and the valley as far up as Newnham assumed the familiar lake-like ap-

pearance which it has borne ever since. Should an elevation
of the land take place in some future age, the former state
of things would be reproduced. A glance at the map will
give an idea of the position of the numerous sand- or mud-
banks formed by the deposits of the tide. We have already
mentioned one of them, "the Noose," in a previous
chapter when describing the bore. The map will also
show the normal current of the river when unaffected by
the tide, and the direction in which it is turned by the
resistance of rocky ground. For example, "Hills Flats"
and "the Ledges" opposite Woolaston give it a sweep to
the right, and from "Beacon Sand" in Tidenham down
to Beachley Point it hugs the western bank. It is a fine
sight from Sedbury cliffs to watch the large ships as they
come up with the tide : for some four miles they keep
close beneath you, and then opposite Pill House they bear
away to the east till you lose sight of them as they enter
the canal at Sharpness.

Were it not for the salmon it would be absurd to talk
of Gloucestershire fisheries, but the Severn salmon is
famous throughout the land. It is less abundant than it
was in past generations, chiefly owing to river pollution
caused by the great towns, but now that the life history
of the fish is better understood and the fisheries are strictly
regulated, it will, it may be hoped, become more plentiful.
The chief method of taking the fish is by nets. The
fisherman moors his boat across the tide, and fixes his net
in such a manner that the bag is carried under the boat by
the current ; he holds a string attached to the net in his
hand and as soon as he feels a fish he hoists up the net.

Another method, which has been practised from the earliest times in the parishes of Tidenham and Woolaston, is that of "potchers," or more correctly "putchers." A long stand of baskets is erected below high-water mark tier over tier; the baskets are long and cone-shaped, and open at the base of the cone. This open end faces the incoming tide, and the fish swimming up the river enter

Salmon "Putchers" at Low Water, Beachley

these baskets, and being unable to turn round owing to the narrow dimensions of the potcher are left high and dry by the receding waters. It need hardly be added that the privilege of erecting potchers has long been limited.

Other famous Gloucestershire fishes are the lamprey and the lampern, which in Gloucester are potted or made into pies. It was formerly the custom for the Corporation to send gifts of these pies to the king, and to certain high officials of the city and county; between 1835 and 1893 the custom was allowed to lapse, but was revived in the latter year. Another fish sold in Gloucester in large numbers in the spring is the "elver". this is merely a very young eel of thread-like appearance, and easily mistaken by the stranger for worms. Millions of these elvers are exhibited in heaps upon the slabs of the fishmongers. Nor must we forget the Woolaston shrimps and Beachley sprats, which always find a ready sale when hawked from door to door by the captors and their families.

Gloucester, before the construction of the Manchester Ship Canal, was the most inland port in the kingdom after London. Its imports are chiefly timber and grain, in which articles its trade is very large: in 1903 the total value of imported goods of all kinds was £2,925,601. Its exports come from the western midlands and from the manufactories in the city itself. In 1904, for example, 30,721 tons of salt from Droitwich and its neighbourhood were exported to foreign ports, and in the same year iron and steel manufactures to the value of £45,579. It will thus be seen that its connexion with Birmingham and other midland towns by railway and canal, together with its easy communication with the Atlantic by means of the ship canal, has made Gloucester the chief town and only port upon the Severn.

Bristol, on the other hand, like Newport and Cardiff on the further shore, is a port of the Channel. But for this city, which has played so important a part in the history of the west, and indeed of England itself, we shall need a separate chapter.

9. The Port of Bristol.

It is to its commerce even more than to its manufactures that Bristol owes its greatness. When Liverpool was unheard of and Cardiff little more than a Norman fortress, Bristol was to the west what London was to the east. Already in the eleventh century she was carrying cargoes of slaves to the Scandinavian settlers—Ostmen as they were called—on the east coast of Ireland, and in the next century permission was received from the king to form a settlement in Dublin. After this the trade with Ireland was no longer confined to slaves. Rough cloth, which was made in Bristol or its neighbourhood, began to be exported not only to Ireland but to Scandinavia, whence the ships brought home cargoes of salt fish—a commodity much more generally consumed in those days than now.

If the map be studied, it will be seen that the site of Bristol is one peculiarly suitable for a trading settlement. Placed on a tidal river capable of carrying ships of large burden, and in easy communication with the ancient town of Bath (and so with all parts of England, for on Bath converged roads from every quarter), merchandise could be collected from, and transported to every part of the kingdom. In the early days when docks were unknown the soft mud of the estuary permitted the careening of vessels without fear of injury. If it is asked why the town was not built nearer the sea, the answer is easy. The flat ground at the mouth of the river was liable to inundation and exposed to the attacks of an enemy by land or sea. The steep cliffs that line the river on both sides made a landing-place impossible till the opening at the foot of Clifton downs was reached. This, then, was the spot, and here quays, alongside which ships could be brought, were constructed. On the land side the neck of the peninsula formed by the Frome and the Avon was easily guarded by a castle, and on this peninsula, fortified land-

The Avon from the Suspension Bridge, Clifton

ward and seaward, the town was built. By nature it was a stronger place than London.

By the time of Henry II, the trade of Bristol had become so important that a charter was granted to the townsmen by the king, freeing Bristol men and their merchandise from the tolls which were then levied upon goods carried through a town or over a bridge. Aquitaine came to Henry as his wife's dowry, and in consequence the trade in French wines received an immense impetus, Bristol being its chief mart. In the thirteenth century the townsmen had what we should call a corporation, and had been granted the same privileges as the citizens of London. Besides this, the various trades had their guilds or societies by which their interests were protected, and the exercise of each particular trade confined to members of its guild.

The fifteenth century, in spite of French and civil wars, was a flourishing time for Bristol merchants. The Merchant Guild had now become the Fellowship of Merchants, a title which itself was to be superseded under Edward VI by that of the Society of Merchant Venturers—a society which is still the wealthiest and most honourable in the city. This was the age of the Canynges, the Cabots, and the Yonges. Their ships sailed north, west, south, and east—to Iceland and the Baltic, to France and Spain, to the Levant, and to the New World. The Canynges—William, John, and William, three generations—seem to have owed their greatness to the cloth trade. The grandson, the greatest of the three, had 2853 tons of shipping afloat and employed 800 sailors: besides these he had at work a hundred carpenters, masons, and other workmen. Thomas Yonge was his half-brother and represented the town in seven Parliaments; Canynges himself was member for Bristol in two Parliaments and was five times mayor. After a long life of prosperity and good works he retired in his old age to the collegiate church of Westbury-on-Trym, where he took priest's orders and died in 1474.

The discovery of the West Indies by Columbus in 1492 had turned the thoughts of Europe to further adventures in the mysterious islands of the west. In such ambition Bristol was the last port to be wanting. In May, 1496, John Cabot with his son Sebastian sailed down the Avon and out into the broad Atlantic. They passed the

Bristol Cathedral

south of Ireland on the right, keeping the head of the vessel steadily westwards. On June 24 they sighted land, probably Cape Breton island between Newfoundland and Nova Scotia ; here they landed and planted the flags of England and St Mark. In May, 1498, Cabot and his son set out again in two ships. One of these proved unseaworthy and was sent home, but in the other Sebastian reached the coast of Labrador. Unable to proceed north-wards on account of the ice he put his vessel about and sailed along the coast southward in search of a passage to the west, for he could not tell whether the land he was coasting was an island or not. He got as far as Virginia, when he gave up and returned to England[1]. For the next hundred years Bristol displayed but little enterprise as regards the New World, though the Spaniards were very active in Mexico and the south. In the seventeenth century more than one attempt at colonisation was made. At last, in 1630, John Winthrop planted a settlement at Charlestown, which was reinforced ere long, and about the same time Robert Aldworth and Giles Elbridge, two Bristol merchants, founded a colony further north in Maine. In 1620 the "Mayflower" had sailed from Plymouth with the Pilgrim Fathers, and by the end of the century Bristol was doing a brisk trade in sugar and tobacco with the southern colonies and the West Indies.

Sugar henceforward became the staple trade of Bristol : the factories in which the sugar was refined and cast into loaves were numerous. Macaulay, describing the wealth of the sugar-refiners in the seventeenth century, says : " The hospitality of the city was widely renowned, and especially the collations with which the sugar-refiners regaled their visitors. The repast was dressed in the furnace, and was accompanied by a rich beverage made of the best Spanish wine, and celebrated over the whole kingdom as Bristol milk." The death-blow to this long-established industry was given by the introduction of Free Trade. The

[1] A tower to the memory of the Cabots now crowns Brandon Hill.

colonial sugar no longer had that advantage in the pay-
ment of duty over foreign sugar, and an inferior article
manufactured in France and elsewhere drove the fine
Bristol sugar out of the market.   The West India planters
who had made their fortunes before this time were considered fortunate.   Another circumstance, which in a great

Old Dutch House and Wine Street, Bristol

degree damaged the West India trade, was the emancipation
of the slaves in 1833.   In the eighteenth century the trade
in negro slaves was one of the chief sources of Bristol
wealth.   Ships laden with home products—iron, cloth,
glass, and so on—sailed to the West Coast of Africa,
where these things were exchanged for a human cargo.
Packed and crowded together in the hold like animals,

and treated in much the same way, the negroes were carried off to the West Indies and sold to the planters. Thence the ships returned home, bringing sugar, rice, tobacco, rum, and other tropical products—soon to be reladen for another voyage. Such was the business of the famous West Indiamen that lined the quays of Bristol, the pride of every citizen and the admiration of every stranger. One would like to believe that among the goods exported in those days were silken fabrics, for many Huguenots had settled in Bristol on the revocation of the Edict of Nantes in 1685, but they seem to have been engaged in weaving woollen materials rather than silk, which employed them in London and elsewhere. They had other trades and professions—a list of 1709 included ten merchants, a physician and three surgeons—and their descendants may be found in Bristol at the present day.

It was in the eighteenth century, too, that Bristol sent out its many privateers. A privateer is a ship-of-war fitted out at the owner's expense, and carrying " letters of marque," which authorise it to engage with any vessel of the enemy, to make prizes, and carry home booty. The most famous of the early privateering expeditions was that of the " Duke " and " Dutchess " fitted out by sixteen Bristol merchants, and commanded by Captain Woodes Rogers. Leaving Kingroad, the great roadstead at the mouth of the Avon, on August 8, 1708, they sailed for Brazil, rounded Cape Horn, found Alexander Selkirk on the island of Juan Fernandez off the coast of Chile, and after taking several prizes set out home across the Pacific on January 1, 1710. They sailed round the

E. G. 4

Cape of Good Hope and reached Bristol with gains to the value of £170,000. Scarcely less successful was one of the cruises of the " Ranger," a Bristol privateer also owned by a member of the Rogers family, which, in 1780, brought into Bristol four prizes, of which one—the "Sta Iñez "—carried treasure only inferior in value to the celebrated Manila galleon "N.S. de Covadonga" captured by Commodore Anson. But the War of the Austrian Succession (1743–48) and the Seven Years' War (1756-63) were the great times for Bristol privateering. At one time no less than fifty-one Bristol privateers were afloat, though often to the loss of their owners.

At the end of the eighteenth century, owing to a combination of causes, the trade of Bristol began to decline. The importation of wool from Ireland had ceased at the end of the previous century when the Irish began to manufacture woollen goods at home, and though this industry was soon crushed by the tyranny of the ruling classes, Bristol gained nothing, for the Irish only took their manufactures to foreign countries. After the wars with Napoleon, the Spanish wool-trade went to London. The difficulty with which the port of Bristol was entered as compared with other ports was now beginning to tell. The want of steam-coal and the heavy export and import dues imposed sent the trade to the north of England. Liverpool had now far outstripped Bristol in the race for trade, and the American ships now preferred to run into the northern port. These heavy port dues, which proved so fatal to the trade of Bristol, were imposed by a Dock Company which had been formed in 1803 to make a harbour $2\frac{1}{2}$ miles long and covering a space of over 80 acres in the centre of the city. This was done by damming up the bed of the Avon, so that its channel formed a long floating harbour, and cutting a new course for the river round the south of the city. The Company, however, cared more about filling their own pockets than fostering the good of the city, and trade kept falling off till, in 1848, the Corporation bought the harbour and reduced the dues.

From this time the trade of Bristol began to revive. Already on April 8, 1838, the Great Western, the first steamship worthy of the name to cross the Atlantic, had sailed down the Avon and reached New York in less than sixteen days, a wonderful achievement at that time, and matters improved slowly till in 1871 a regular trade with America was once more established. In 1877 a new dock was opened at the mouth of the Avon by an independent Company, and in 1880 another Company opened a dock on the other side of the Avon at Portishead two miles west of the former dock. A railway on either side of the river connected these docks with the city. The competition which sprang up between these new docks and the old city docks proved disastrous to the latter, and in 1884 they both passed into the hands of the Corporation.

The trade of Bristol now began to prosper, as it had never done since the days of steam ships, and the increase in the size of the great ocean steamers made it clear that if Bristol was to hold its own further dock accommodation would be necessary. A new dock, called the Royal

Edward Dock, was therefore begun in 1902 on the outer side of the Avonmouth dock, which was connected with it by a canal 85 feet wide. The new dock, which covers 30 acres, was opened by King Edward VII in July 1908[1].

Avonmouth Docks

10. Climate.

The climate of a country or district is, briefly, the average weather of that country or district, and it depends upon various factors, all mutually interacting; upon the

[1] The following table will show the size of the various docks

City Docks	.	.	.	.	83	acres
Avonmouth Dock	.	.	.	19	„
Portishead Dock	.	.	.	12	„
Royal Edward Dock .	.	.	30	„

All are under the control of the City.

latitude, the temperature, the direction and strength of the winds, the rainfall, the character of the soil, and the proximity of the district to the sea.

The differences in the climates of the world depend mainly upon latitude, but a scarcely less important factor is this proximity to the sea. Along any great climatic zone there will be found variations in proportion to this proximity, the extremes being "continental" climates in the centres of continents far from the oceans, and "insular" climates in small tracts surrounded by sea. Continental climates show great differences in seasonal temperatures, the winters tending to be unusually cold and the summers unusually warm, while the climate of insular tracts is characterised by equableness and also by greater dampness. Great Britain possesses, by reason of its position, a temperate insular climate, but its average annual temperature is much higher than could be expected from its latitude. The prevalent south-westerly winds cause a drift of the surface-waters of the Atlantic towards our shores, and this warm water current, which we know as the Gulf Stream, is the chief cause of the mildness of our winters.

Most of our weather comes to us from the Atlantic. It would be impossible here within the limits of a short chapter to discuss fully the causes which affect or control weather changes. It must suffice to say that the conditions are in the main either cyclonic or anticyclonic, which terms may be best explained, perhaps, by comparing the air currents to a stream of water. In a stream a chain of eddies may often be seen fringing the more steadily moving central water. Regarding the general north-easterly moving air from the Atlantic as such a stream, a chain of eddies may be developed in a belt parallel with its general direction. This belt of eddies or cyclones, as they are termed, tends to shift its position, sometimes passing over our islands, sometimes to the north or south of them, and it is to this shifting that most of our weather changes are due. Cyclonic conditions are associated with a greater or less amount of atmospheric disturbance; anticyclonic with calms.

The prevalent Atlantic winds largely affect our island in another way, namely in its rainfall. The air, heavily laden with moisture from its passage over the ocean, meets with elevated land-tracts directly it reaches our shores—the moorland of Devon and Cornwall, the Welsh mountains, or the fells of Cumberland and Westmorland —and blowing up the rising land-surface, parts with this moisture as rain. To how great an extent this occurs is best seen by reference to the accompanying map of the annual rainfall of England, where it will at once be noticed that the heaviest fall is in the west, and that it decreases with remarkable regularity until the least fall is reached on our eastern shores. Thus in 1906, the maximum rainfall for the year occurred at Glaslyn in the Snowdon district, where 205 inches of rain fell; and the lowest was at Boyton in Suffolk, with a record of just under 20 inches. These western highlands, therefore, may not inaptly be compared to an umbrella, sheltering the country further eastward from the rain.

The above causes, then, are those mainly concerned

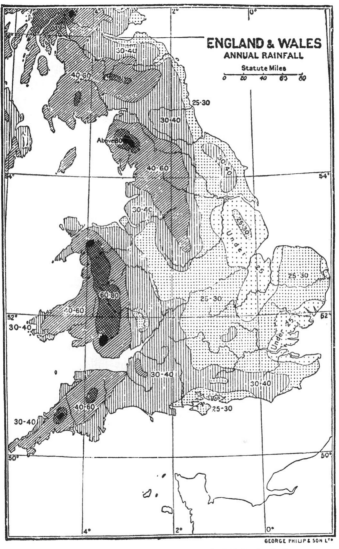

(The figures show the annual rainfall in inches.)

in influencing the weather, but there are other and more local factors which often affect greatly the climate of a place, such, for example, as configuration, position, and soil. The shelter of a range of hills, a southern aspect, a sandy soil, will thus produce conditions which may differ greatly from those of a place—perhaps at no great distance—situated on a wind-swept northern slope with a cold clay soil.

The character of the climate of a country or district influences, as everyone knows, both the cultivation of the soil and the products which it yields, and thus indirectly as well as directly exercises a profound effect upon Man. The banana-nourished dweller in a tropical island who has but to "tickle the earth with a hoe for it to laugh a harvest" is of different fibre morally and physically from the inhabitant of northern climes who wins a scanty subsistence from the land at the expense of unremitting toil. These are extremes; but even within the limits of a county, perhaps, similar if smaller differences may be noted, and the man of the plain or the valley is often distinct in type from his fellow of the hills.

Very minute records of the climate of our island are kept at numerous stations throughout the country, relating to the temperature, rainfall, force and direction of the wind, hours of sunshine, cloud conditions, and so forth, and are duly collected, tabulated, and averaged by the Meteorological Society. From these we are able to compare and contrast the climatic differences in various parts. In Gloucestershire, as may be guessed, the chief distinction is between the Cotswolds and the Vale, and the protection, good soil, and other advantages which the latter enjoys have a marked effect upon its productions. "As long coming as Cotswold barley" is an old saying referring to this fact, and the ripening of the crops on the hills is often a month later than in the Vale.

In the rainfall map it will be seen that Gloucestershire lies just outside the limits of the 40—60 inches average annual fall and is placed in the 30—40 inches division. There are, however, exceptions to this, and a small isolated region of the heavier fall exists just north of Bristol. At Clifton the rainfall for the thirty years from 1870 varied between 35 and 45 inches, and at Stroud for the same period between 29 and 85 inches. These figures are considerably higher than the average annual rainfall for Great Britain, which is about 32 inches. On the whole the south of the county is wetter than the north.

On the whole Gloucestershire is a warm county, partly due to the fact that the temperature of the water in winter is warmer than that of the land, and to the warm south-westerly winds that blow up the Bristol Channel. The mean annual temperature for most of the county is between 48 and 50 Fahr., but in the valley of the Bristol Avon and of the Severn as far up as Gloucester it is over 50 Fahr. Partly owing to its mineral springs and partly to its sheltered position Cheltenham has become a favourite health-resort, especially suitable for the old and delicate.

11. People—Dialect, Settlements, Population.

Although Gloucestershire belonged to Mercia for a great part of its early history, its people were West-Saxon. They had subdued the Hwiccas, as the old British settlers were called, who by that time—at least in the towns—had become more or less Romanised. There were at the time of the Conquest two dialects spoken in England, the northern and the southern, and the men of Gloucester of course spoke the latter. But with the Normans a multitude of French words came in, and helped to break up the old division. Of all these modified forms it was the east-midland which became the language of London and of literature, and the pure southern dialect was pushed westwards till it deserved the name of western rather than southern. This western form remained the language of Gloucestershire, Wiltshire, Berkshire, Somerset, Dorset, and Devon. As language is not made but grows, further modifications arose, till at the present day the man of Gloucestershire complacently tolerates what he considers the curious dialect of the man of Devon. The old Gloucestershire dialect in its purity may be found in the Chronicle of Robert of Gloucester.

Gloucestershire has been singularly free from foreign settlements: the only noteworthy immigrants being the French refugees of the sixteenth and seventeenth centuries. It was they who introduced the art of silk-weaving into the Stroud valley, and at France Lynch near Chalford a French inscription may still be seen on one of the cottage windows.

The population of the county in 1921 was 757,668 and the number of square miles is 1259, which gives an average of 602 to the square mile. Gloucestershire is the seventeenth English county in size, and the thirteenth in population.

12.   Agriculture—Cultivations, Stock.

Few counties have so many rural industries exclusive of agriculture. In the whole county including the towns 60,000 persons are engaged in various industries, while the number of those engaged in agriculture is but 25,000. The total number of acres under crops and grass in 1923 was 626,828 ; of these the corn crops, including rye and mixed corn, covered 93,148 acres, and 401,307 acres were laid down in permanent pasture, the rest being taken up with green crops, clover and sanfoin, etc.

The general features of the county have been described in a previous chapter, we have now to consider them from the point of view of their products. At the present day, though here and there many acres have reverted to the original sheep run, the general surface of the Cotswolds is covered with cereals and artificial herbage. The soil is a calcareous loam three to six inches deep, locally called stonebrash : the ploughing must therefore be very shallow, and rolling as well as treading by sheep is important. For fodder the characteristic plant, as before noticed, is the sainfoin. " This princely herb," as an eighteenth century

writer on the agriculture of the county styles it, is a native,
but it has been cultivated on these hills for some 250
years, and was a familiar plant long before the cultivation
of the turnip was introduced. The fact that its roots
penetrate the interstices of the oolite rock beneath the
stonebrash makes it a crop peculiarly suited to this shallow
soil, and about 15 per cent. of the arable land is devoted

The Agricultural College, Cirencester

to it. The crop stands from three to five years before it
is broken up. To the south-east of the hills round
Fairford, South Cerney, and Tetbury the soil is clayey,
more or less overlaid with gravel, the same as that of the
adjacent county of Wilts. Since the disastrous years, i.e.
since 1879, the upkeep of the untractable arable land in
this district has not been worth the expense, and much of

it has been laid down in pasture. Dairy farming and the supply of milk to London and Swindon is now the chief industry. The water meadows along the lower reaches of the Churn and Coln are carefully irrigated, and produce at least two rich cuttings of grass in the season.

The Severn valley consists chiefly of Lias clay overlaid in some places by a sandy loam and in others by a red marl, said to be brought down from the red country in Worcestershire by the Teme, a tributary which joins the Severn just below Worcester. In the lower parts nearer the Bristol coalfield the Old and New Red Sandstone occur. The whole vale is extremely fertile, and half a century ago the arable and pasture land was about equal, but now the greater part is pasture. Besides dairy farming and the manufacture of butter and the celebrated Double and Single Gloucester cheese, store cattle are reared for the market in large numbers, and in many of the parishes the famous Gloucestershire cider has been made for many generations. The earliest mention of it which has been discovered is across the river at Tidenham in about 1290, but no doubt it had even then been made there for some centuries. In the vale of Evesham not only pears and apples but other fruits are extensively cultivated, and near the large towns—Cheltenham, Gloucester, and Bristol— there are many acres of market-gardens.

Out of 73,000 acres west of the Severn, some 24,000 are forest. Of the rest the Newent district, as we have called it, consists of a light loam lying on the Old Red Sandstone. It is by no means a rich or very fertile soil, but will grow barley and turnips well, and feed cattle and

A Cotswold Homestead

sheep. These sandy soils have long been known as the
Ryelands, and gave their name to a breed of small sheep.
The "Ryelanders" however, having been continually
crossed with other breeds to improve their size, are very
seldom found in a pure state: they are a small, white-
faced hornless breed, remarkable for the fineness of their
wool and for the excellence of their mutton.

The Old Red Sandstone and Carboniferous Limestone
continue southwards from the Forest to the extremity of
the peninsula: near the Severn a considerable area is upon
New Red Sandstone, Red Marl and Lias. In the sandstone
parts, which are considerably in excess of the limestone,
the cultivation resembles that on the eastern side of the
estuary. The depth of soil on the limestone rock is
small and seldom worth more than rough grazing for sheep
and cattle.

The number of cattle have increased with the increase
of pasturage; about Tewkesbury there are a few Here-
fords, but the Shorthorns are almost universally kept. Of
the old Gloucestershire breed only a very few specially
protected herds survive, as at Badminton and Prinknash.
A writer at the beginning of the last century says that
even then few "packs" of the genuine old stock were
preserved. He describes them as differing "little in general
appearances from the Glamorganshire, except in colour,
which is a dark red, or brown: the bones are fine, horns
of a middling length, white in colour, with a black tip at
the ends...a slight streak of white generally runs along
the back, and always on the rump end of the tail."
Oxen are still employed for the plough on the Cotswolds,

Windrush, Oxen ploughing

though in some parts they are now almost entirely disused. This is to be regretted, especially on large farms which can keep a team of horses as well, as the expense of their keep is only two-thirds that of the horses, and they can do nearly as much work, both at the plough and as draught animals. Harnessed to the ox-cart or wain they formerly did all the farm traffic now done by horses.

The famous Cotswold sheep, which for many generations made Chipping Campden and Northleach the centre of the wool-trade of England, are now very scarce. In the latter part of the eighteenth century they had already been crossed with the Leicesters, and since then, mutton having become a more important commodity than wool, they have been largely crossed with the South Down, which has reduced their size, and also the quantity of the wool. The demand for large joints of mutton is not what it was, while for small joints the butchers always find a ready sale. A few flocks of the pure breed, however, are still left. Those at Daglingworth and Hampnett were, and may be still, famous; and in 1905 the celebrated Aldsworth flock produced 1100 fleeces at an average of 10½ lbs. per fleece. More recently, too, the breed has been revived at Hyde Farm, near Stroud.

Gloucestershire is not a great horse-breeding county, though most farmers have a foal or two in the season.

The causes of agricultural depression were the same in Gloucestershire as in other counties, viz. the long run of bad seasons culminating in 1879, and the importation of foreign cereals. The result of the latter in our county is shown at once by the fact that between

the years 1879 and 1909 permanent pasture increased by
114,639 acres.   The decrease in the number of agricultural labourers is due partly to the same cause, and
partly to the introduction of machinery, fewer hands
being needed in the work. Yet the flail may still be seen
in some districts, or might have been not long ago.   There

Threshing with the Flail

is, however, a growing tendency of the young people of
both sexes to flock to the towns.   In 1871 the number
of labourers in the county was 20,586, in 1901 13,319:
in the former year 2007 of these were women, in 1901
only 182.

Still, after all, Gloucestershire has not suffered so much

as many other counties; there is a spirited competition for farms, and though farmers are a race privileged to grumble, it is seldom that a man of industry and prudence fails to make a decent living.

13. Industries.

The staple industry of the county during the greater part of its history was wool. For six hundred years the Cotswolds were a vast open down, over which roamed thousands of sheep whose wool was famous throughout England. When shorn the wool was stored in great warehouses at the wool centres, of which Cirencester, Northleach, and Chipping Campden were the chief. The London wool-merchants then came down, and purchased the wool from the middlemen by the sack and "tod." They then packed it in "sarplers" and "pokes" and conveyed it to the Kentish ports, whence it was shipped to Calais or to some Flemish wool-mart. This went on all through the Middle Ages, but under Edward III the Gloucestershire men not only grew the wool but also began to weave it into cloth. For this purpose water was necessary, and the Stroud valley having the best and swiftest streams, it was here that the looms were set up. The wool could now be washed and dyed, and in later days water-power machinery could be set up. This was the origin of the Gloucestershire clothing trade, which flourished down to the middle of the nineteenth century. Its history cannot be given here even in outline; it must suffice to say that the chief causes of its decline were, the introduction of machinery, which employed fewer hands and so met with

violent opposition; the rise in the price of the fine foreign
wools partly owing to the eight years' drought in Australia;
and lastly the high protective tariffs of foreign countries
combined with their absence at home. These are some of
the difficulties with which the great manufacturing towns
of the north have been so far able to cope, but which have

Chipping Campden Market House and old Guild Chapel
now the Town Hall

given a severe blow to the clothing trade of the west.   It
was in 1830 that the decline began, and since that time
the mills at Stroud are the only ones that have maintained
their ground.   Here of late years there have been distinct
signs of improvement.   A certain number of firms have
always kept their mills open, and a method has now been

discovered of refining the Cotswold wool and weaving tweeds from it as good as any of those of the north of England. Not long ago it was considered too coarse for any but the rougher kind of material. Now, owing to the improvements and the advance in the construction of machinery, the annual output of Stroud cloth is larger than it ever was.

If the number of persons engaged gives a county the right to be called "industrial," Gloucestershire may be so called, for, as we have seen, 60,000 persons are engaged in industries of one sort or another, and only 25,000 in agriculture. The former number of course includes persons engaged in mining, which is the subject of the next chapter. Many are employed in the Woods and Forests, which cover about 61,184 acres—an area only surpassed by five other English counties, namely York-shire, Hampshire, Sussex, Kent, and Devon. Timber-dealing employs nearly 7000 persons, for, as noted in a former chapter, there is a considerable trade with Norway and other countries in this article. In 1905 £25,000,000 worth was imported by the whole country, of which Bristol was responsible for £1,000,000, and Gloucester for £600,000. In the days when our navy consisted of wooden ships, the Forest was one of the chief sources from which the timber was drawn. There are 15,000 acres now under timber cultivation, and this con-sists of young oaks. The ancient trees were mentioned in a former chapter.

Other trades which employ a large number of hands are—engineering machinery and shipbuilding 10,470, and

Morris Dancers, Chipping Campden

miscellaneous iron and steel, 2832. Bristol is now the place where most of the Great Western locomotives are built, with the exception of the heavy main-line engines, which are built at Swindon. Other manufactures for which Bristol is known all over the world are those of cocoa and tobacco, the former having been in existence for 180 years. To agricultural manufactures we referred in the last chapter. There are many other smaller industries of which we cannot speak here, but we must not forget bell-founding, for which Rudhall of Gloucester was famous two hundred years ago. The industry was revived at Bristol in 1875 and is now a flourishing trade.

14. Mines.

The principal mineral products of Gloucestershire are coal and iron. There are two coalfields—the Forest and that part of the Bristol coalfield which lies between the Avon and Tortworth. In 1903 1,406,021 tons of coal were raised from these two fields. Taking the latter first, its area is about 50 square miles, or ten miles long by five wide, and there are now less than 30,000,000 tons of the best coal to be worked. The output is about 250,000 tons a year and at this rate it will be exhausted in about another 100 years. This affords the principal household coal of Bristol.

On the other hand the area of the Forest coalfields is about 34 square miles and its annual output is more than

460,000 tons, at which rate it is calculated that the supply will last for another 600 years. Firedamp in these pits is unknown, and the miners work with naked lights. A well-known Gloucester man writing 20 years ago says: "One can hardly imagine anything more severe in the way of labour than that of the miner, lying on his side in

Speech House Colliery

a four-foot passage, cutting away with his pick the hard rock encasing the seam. The strain upon the hands, wrists, and arms is tremendous. The swing of the body is not available, the legs are miserably cramped away, and the whole task seems to fall upon the arms; but use is everything. The men do not feel the cramped position, and prefer this work, to which they have been born and bred, to any other. 'Nothing tires me like having to stand long on my legs. The worst of all is gardening. Bending down like that breaks my back,' said an old hand who had been 52 years a pitman. This dislike of gardening is common among colliers in the Forest of Dean."

Geologically speaking the coal lies in a basin, of which the inner coat is Millstone Grit, the middle coat Carboniferous Limestone, and the outer coat Old Red Sandstone. A writer of the middle of the last century says, "The coal-measures in the central parts of Dean Forest are 1410 feet deep....There are 27 beds of coal, making an aggregate thickness of 40 ft. 8 in. These seams vary from one or two inches to two or three feet, and in one instance—Coleford High Delf—[one of the lowest seams] to five feet. The Millstone Grit is from 265 to 455 feet thick."

The head officer under the Crown who has the control of the mines is called the gaveller, and by ancient custom men born in the Hundred of St Briavels, who have resided in the said Hundred, and worked in the mines for a year and a day, have the privilege of being registered as "free miners." Only free miners have a right to grants of the Crown mines.

The iron ore is found in the Millstone Grit and in the upper part of the Limestone, and has been worked from the days of the Romans. They worked underground, and also by what are called "day-workings," that is deep winding passages open at the top, and now called "Scowles." But the ore they got out they only partially smelted, and the unexhausted "cinders" which they left were profitably reworked at least as early as the time of Edward I. The fuel used for smelting up to the time of Dud Dudley, an ironmaster of the days of Charles I, was oak and beech charcoal; for although "sea-coal" or "pit-coal" was worked as early as the thirteenth century it was not thought possible to utilise it for smelting purposes till the appearance of Dudley's book *Metallum Martis*, in which he suggested the experiment. This took a long time to bring to perfection, and when at last this was accomplished, the Sussex ironmasters whose supply of timber was getting exhausted migrated to South Wales for the sake of the coal.

The supply of iron ore is, however, on the decline; for in 1856 the total yield of all the iron mines in the Forest was 109,268 tons, and in 1903 it was only 7456 tons.

Iron was also worked at a very early period in the Bristol coalfield, where there is a village called Iron Acton.

15. History of the County.

At the earliest period which need be noticed in this chapter the district afterwards called Gloucestershire was peopled by a British tribe known to the Romans as the Dobuni. On the invasion of the Romans, this tribe was subdued in A.D. 43 by the generals Aulus Plautius and Ostorius Scapula. The Romans founded at Gloucester (Glevum) one of the four colonies which they planted in

the island, the other three being Colchester, Lincoln, and
York; and they established a settlement for residence and
commerce at Cirencester (Corinium).   Apart from these
two towns, villas and smaller settlements, as we shall see
in the next chapter, were scattered over the Cotswolds.

On the withdrawal of the Roman troops about 410,
Gloucestershire, to use the later name, fell back under

Hypocaust of Roman Villa, Chedworth

native rulers.   By this time the upper classes had become
Romanised, and many of them were Christians.   In
towns like Cirencester they lived in ease and comfort, and
the civilisation equalled that of the continental towns,
while the wealthier citizens had their villas in convenient
situations in the country.   In Gloucestershire probably

the inroads of the Picts and Scots on the north-west, and
of the marauding Saxons in the south-east, were little
heeded, but the crash came when in 577 the West-Saxon
princes Ceawalin and Cutha crossed the lower Cotswolds
and defeated at Dyrham the three "kings" of Bath,
Gloucester and Cirencester.   Thus the whole of the
county (to use the word in anticipation) east of the Severn

The Abbey Gate, Cirencester

became Saxon, and the Welsh that survived were either
reduced to slavery or driven beyond the river. The Saxon
tribe which finally settled in the subsequent counties of
Gloucester and Worcester was the Hwiccas, and this
tribe formed the diocese of Worcester, when the great
Mercian diocese of Lichfield was divided in 673. It was
under the rule of the Mercian kings that the great dyke,

commonly attributed to Offa, was thrown up from the Dee to the Severn to mark the boundary between the English and the Welsh. The southern part of this dyke may still be traced in the west of our county running along the summit of the cliffs on the left bank of the Wye. Its termination on the Severn is conspicuous at Sedbury in Tidenham. The province of the Hwiccas was not divided into the shires of Worcester and Gloucester until Mercia was recovered from the Danes in the tenth century. The old kingdom of Mercia was then split up into divisions called shires which took their name from the chief town of each district, and thus differed from the counties of the north, east, and south, the boundaries of which corresponded with those of the old English kingdoms. These latter counties, it will be noticed, are not called after their capital towns.

In the autumn of 1066 came the Norman invasion, and a year and a half later Gloucestershire and the adjacent counties were subdued. Gloucester, which under the Confessor had taken the place of Oxford as a central city for assemblies and councils, retained its dignity under the conqueror who "wore his crown" here at the Christmas assembling of the Court, as he did in Winchester at Easter, and in Westminster at Whitsuntide. The only other event which need be mentioned before the Wars of the Roses was the foul murder of Edward II at Berkeley Castle (1327), nine months after his deposition.

The battle of Tewkesbury is one of the famous battles of English history. In the spring of 1471 Edward IV landed in Yorkshire after his temporary deposition and exile. On April 14 he crushed the Earl of Warwick and the Lancastrians at Barnet. On the same day Queen Margaret, who had been in France ever since her defeat at Hexham in 1464, landed at Weymouth with the intention of joining Warwick. She now changed

Berkeley Church and Bell Tower

her plans and was marching northwards to join Jasper Tudor in Wales, when Edward came up with her at Tewkesbury. The Lancastrians were drawn up on the south side of the town, and early on the morning of the fourth of May, before they could be reinforced by Jasper Tudor, who was marching from Chepstow to their assistance, the battle began. The Lancastrians were utterly routed, their leaders slain and Queen Margaret taken prisoner. Edward was now firmly established on the throne.

The dissolution of the monasteries, and the enormous

The Bear Inn, Tewkesbury

change which it produced in the whole face of the country, is part of the general history of the nation, but in few counties can the shock have been more widely felt than in Gloucestershire, with its wealthy abbeys and numerous smaller religious foundations. It is certain that the farmers did not benefit by the change, for they exchanged good landlords for a horde of greedy adventurers, who cared nothing for farming beyond the rents.

There was no corner of Gloucestershire that did not feel the stress of the civil wars of the seventeenth century. The constant marching to and fro of troopers and dragoons, musketeers and pikemen, and the frequent skirmishes between the two parties, left the inhabitants

The Bell Inn, Tewkesbury

no rest. The billeting of soldiers and the requisitioning of provisions spared neither friend nor foe. The townsmen of every defensible town either had to stand the horrors of a siege, or to put themselves at the mercy of the challengers. The farmer had to see his land run to waste, while his horses and oxen were hauling the great guns up the hills, or pressed into the service of the baggage train. But the most famous event, as well as that which had most influence upon the future course of the war, was the resistance of the city of Gloucester to the attack of the Royalists. By the middle of 1643 the whole county with the exception of the capital city was in the hands of the king. If he was to be master of the country beyond

Market Cross, Stow-on-the-Wold

the Severn, and if he was to march upon London, he must first capture this stronghold of Puritanism. On August 10, he sat down before the city in person. The citizens under Colonel Massey held out for four weeks, and on September 5 the Earl of Essex with an army of relief appeared upon the edge of the Cotswolds and the siege was at an end.

We have no space to mention any of the other affairs in our county except the battle of Stow, March 21, 1646, the last encounter of the war in the open field. Here the veteran Lord Astley, the hero of many fights, laid down his arms to the Roundheads. With clear foresight of what was to come, he told his opponents that they might now sit down and play, since their work was done, if they did not fall out among themselves.

At the Revolution the Lord Lieutenant, the first Duke of Beaufort, tried to hold the county for King James, but the Prince of Orange was soon in a position to make all resistance useless.

In the eighteenth century fighting was exchanged for electioneering squabbles, and in the nineteenth for rioting —firstly in opposition to the introduction of machinery in the clothing trade, and secondly in favour of the Reform Bill. It was in the latter connexion that the famous Bristol riots of October 1831 occurred.

The domestic history of the county is on the same lines as that of other English counties, that is, in the country it tells of land tenure and the cultivation of the soil, in the towns of trade and industries. In the middle of the fourteenth century Gloucestershire, like the rest of England, was swept by the terrible scourge known as the Black Death, which carried off half the population. The middle of the fifteenth century saw the beginning of enclosures, and the practice of sheep-farming on a large scale ; the eighteenth century brought the destruction of the old system of open-field cultivation and the extinction of the yeoman farmer. We have already set down the population to the square mile of the county at the present time; after the Black Death it was only from 40 to 60, a proportion which was also that of Kent and Essex. But it would take a much bigger book than this to write the domestic history of the county. In this chapter we must be content with the notice of the few occasions on which Gloucestershire became the scene of events which form a part of the history of the whole kingdom.

16.   Antiquities—Prehistoric, Roman.

When we talk of the prehistoric period in the history of a country we mean the period before the commencement of written records. The date at which the historic period begins naturally differs in different nations. In our country it may be fixed at the invasion of Julius Caesar, 55–54 B.C. But the absence of written history does not mean that the prehistoric period is utterly unknown to us. The men of those far remote days have left traces of themselves and their work both in the remains of their dwelling-places and in their tombs, and we find their weapons, and their domestic implements. These it is the business of the antiquarian, as distinct from the historian, to study and interpret. The results of such investigations have led to the dividing of the life of man on our island into three periods, based upon the material used for the manufacture of weapons and domestic implements: (1) The Stone Age; (2) The Bronze Age; (3) The Iron Age; but it must be remembered that these terms are independent of dates, that the periods overlapped, and

Palaeolithic Flint Implement
(From Kent's Cavern, Torquay)

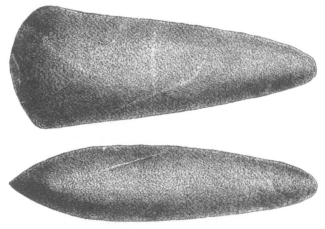

Neolithic Celt of Greenstone
(From Bridlington, Yorks.)

that they differed so much in different countries that even at the present day we have some parts of the world, e.g. New Guinea, which are still in the Stone Age. With us, however, the Stone Age is divided into two periods separated by a considerable period of time, the old or palaeolithic, and the new or neolithic. The men of the palaeolithic age lived in holes and caves, and their implements were roughly-chipped flints. Our island was then connected with the continent of Europe, and huge beasts such as the mammoth and the woolly rhinoceros came to and fro. Their bones, together with the flint implements of palaeolithic man, are found in the deposits brought down by ancient rivers, and under the limestone floors of caves.

A long gap, it is thought, occurred between palaeolithic man and the coming of his neolithic successor in Britain. The latter was probably of Iberian race, the present form of which is now represented by the Basques in the north-west of Spain, while survivals are also found in the short dark-haired inhabitants of South Wales. The neolithic implements are the earliest we find polished and sharpened, though they exist in other forms and need not be further described here. Neolithic man kept cattle, and grew wheat. He lived in huts hollowed out in the ground, and roofed with turf or wattle. On the approach of an enemy he moved his family and cattle up to one of the "camps" on the nearest eminence. These camps are very numerous on the top of the escarpment of the Cotswolds and elsewhere, and were afterwards often utilised and modified by the Celt, the Roman, and the Saxon in turn. The neolithic men buried their dead in the long barrows which may still be seen here and there on the hills. Even palaeolithic man made rude scratches of animals on bone, and neolithic man fashioned for himself rude vessels of baked clay with some kind of a pattern upon them.

Wagborough Bush Barrow, Bourton-on-the-Water

Long after the Iberians came the Celts, an Aryan people, dark and round-headed, who made their weapons of bronze, and buried their dead in round barrows. Sometimes these barrows were mere tumps, at others they were made of large stones forming a sort of hut inside which the body was placed, and the whole then covered with a mound ; these are now distinguished as dolmens.

We have many barrows, but no dolmens left in Gloucestershire. These Celts came over in two divisions. The first, called the Goidels, drove the stone men westwards before them and entered into their inheritance, only to be themselves driven in the same direction by the second horde of Celts, the Brythons or Britons. The Goidels were pushed north-westwards to Scotland and Ireland, just as the Britons were pushed into Wales and Cornwall when the Saxons came.

As for the Romans, their policy was not one of expulsion and extermination; they subdued the Britons and mingled with the conquered tribes; they introduced among them the civilisation of Rome with so much success that all the wealthier classes and the population of the towns became entirely Romanised. The traces of the Roman occupation in our county are abundant. "Villas" or the country houses of the upper classes are scattered all over the Cotswolds, though most of them still have to be unearthed. The most famous are those at Chedworth and Woodchester. The Roman towns of Gloucester and Cirencester are buried beneath the later towns. After their destruction by the Saxons they long lay waste, and though the four main streets which lay north, south, east and west from the Cross at Gloucester no doubt follow the four ways of the Roman castra, at Cirencester the town lay in ruins for so long that when rebuilt it was laid down on entirely different lines. The Roman town has therefore only been excavated here and there, but above ground the amphitheatre may still be seen, and on the east side of the town a portion of the Roman wall.

Of the ancient roads, some belong to the Celtic period, others were constructed by the Romans. Where a great Celtic track existed the Romans would utilise it, repairing and remaking it where necessary. This was the case with the most famous of all the Gloucestershire roads, the great Fosseway, which as we shall see still forms the main road from Cirencester to Moreton and beyond.

From what has been said it will be seen that the whole of the county (not excepting the Forest, in which the Romans had their ironworks) was more or less Romanised, but the point to notice is that the Roman rule was one of peace. The colony of Gloucester (Glevum) no doubt contained a small garrison, but like Cirencester it was mainly a seat of commerce and manufacture. In fact both places served the purpose of market towns as we know them to-day.

17. Architecture — (a) Ecclesiastical. Churches, Cathedrals, Abbeys.

The religious houses in Gloucestershire and the number of churches which they, as well as other foundations elsewhere, were allowed to appropriate, may be gauged by the number of vicarages, as distinct from rectories, at the present day. Let us set down the Orders which had property in the county, together with their houses. The Benedictines had Deerhurst Priory, St Peter's Abbey, at Gloucester, Winchcombe Abbey, Tewkesbury Abbey, and Stanley St Leonard's Priory. The Cistercians had

Hayles Abbey and Flaxley Abbey. St Augustine's Abbey, Bristol, and Cirencester Abbey, together with the Priories of St Oswald and New Llanthony at Gloucester, were filled by Augustine or Austin Canons. The Cistercian Abbey of Kingswood was till the last century an island of Wiltshire. Besides these there was a College of Secular (not Regular as at the Augustine houses) Canons at Westbury-on-Trym, and a Preceptory (as the houses of the military orders were called) at Quenington. The Templars had lands at Temple Guiting. After the Dissolution all these monastic houses were pulled down and carted away, but the churches of three of them were spared—those of St Peter's, Gloucester, and St Austin's, Bristol, to serve as cathedrals for the new dioceses created by Henry VIII, and that of Tewkesbury for use as the parish church.

Before these new dioceses were formed the whole of the country east of the Severn and north of the Forest peninsula had belonged to the diocese of Worcester. The Forest peninsula south of a line drawn north-westwards from Minsterworth to Lea had been in the diocese of Hereford, but, in 1541, this, together with the rest of the county, was taken to form the new diocese of Gloucester, and next year a small portion in the extreme south was cut off to form, with Dorset, the new diocese of Bristol.

We will consider the architecture of the buildings in Gloucestershire under three divisions, viz.: (*a*) Ecclesiastical, or buildings relating to the Church; (*b*) Military, or Castles; (*c*) Domestic, or houses and cottages.

There is one fact worth noting with regard to all

these classes of buildings, and that is that—as indeed is universally the rule—the architecture of the county has been affected by the materials accessible. Thus we find that stone, wood, flints, and bricks are used either because

The Font, Deerhurst Church

they could be easily obtained, or because of the wealth or otherwise of the builders.

Now with regard to the ecclesiastical buildings, let us consider first the churches and cathedrals, and then glance at the remains of the abbeys, monasteries, and other religious houses. The churches of Gloucestershire are

Norman Doorway, Lower Guiting

of various styles and of different ages, so that it will be well to classify them as Saxon, Norman, Early English, Decorated, and Perpendicular.

Towards the end of the twelfth century the round arches and heavy columns of Norman work began gradually to give place to the pointed arch and lighter style of the first period of Gothic architecture which we know as Early English, conspicuous for its long narrow windows, and leading in its turn by a transitional period into the highest development of Gothic—the Decorated period. This, in England, prevailed throughout the greater part of the fourteenth century, and was particularly characterised by its window tracery. The Perpendicular, which, as its name implies, is remarkable for the perpendicular arrangement of the tracery, and also for the flattened gateway arches and the square arrangement of the mouldings over them, was the last of the Gothic styles. It arose somewhat suddenly after the Black Death and was in use till about the middle of the sixteenth century. It is confined to our country, and unknown on the Continent.

We will begin our brief survey of the various styles of ecclesiastical architecture with the two cathedrals. Gloucester was the church of the abbey of St Peter. Very little remains of the conventual buildings except the cloister ; this is on the north side of the church, and is famous as the earliest example of the peculiarly English style of vaulting known as fan-tracery. On the south side are the "Carrells," i.e. window recesses in which the monks read and wrote. On the north side is a long trough in which the monks washed before meals, with a recess for towels opposite. The cloister served as a means of communication between the church and other parts of the buildings, and also for exercise in wet weather. The nave is late eleventh century work, the aisles communicating with the nave by semi-circular arches springing from massive round pillars. The choir was originally Norman also, but in the fourteenth century a Perpendicular appearance was given to the whole of the interior. This, however, does not prevent the original Norman work from being detected at the back of the triforium.

Cloisters and " Carrells," Gloucester Cathedral

The Lady Chapel is late Perpendicular. The most famous tombs are those of Edward II, whose murdered body was brought here from Berkeley Castle (1327), and of Robert Duke of Normandy, the eldest son of the Conqueror, who died a prisoner at Cardiff Castle in 1135.

Bristol Cathedral was the church of St Augustine's Abbey. Founded by Robert Fitzharding in 1142, it consisted of the usual conventual buildings, of which little but

Gloucester Cathedral, East End of Choir

the fine Norman gateway and a part of the cloister remains. It was surrendered to Henry's commissioners in December, 1539, and the first bishop appointed was Paul Bush, a canon of Salisbury.   At this time the church consisted of a choir and north and south transepts : the nave was added in 1877.   The choir is not so beautiful as at Gloucester, but is remarkable for the large Decorated

The Chapter House, Bristol Cathedral

aisle windows, and the fourteenth century glass of the large east window.   Bishop Butler, the author of the *Analogy*, is buried in the cathedral (1752).   Among the other monuments are some to early members of the Berkeley family, including that of the founder.

The church of Tewkesbury Abbey, though now only a parish church, may rank with any cathedral in the land for size and beauty.   Founded in 1105 by Robert Fitz-

Tewkesbury Abbey, the Nave looking East

Hamon, Earl of Gloucester, it was one of the chief monasteries in the kingdom, and its abbot sat in Parliament.   The most striking feature of the exterior is the grand round-headed and recessed arch at the west entrance, 64 feet in height, now filled with a late Perpendicular window.   In the interior the great Norman nave with its huge round pillars reminds one at once of Gloucester. The transepts and the tower are also Norman, but the choir, like that at Gloucester, has been altered, though in the Decorated instead of the Perpendicular style.   The eastern extremity of the church is hexagonal, and here are many splendid tombs and monuments, including that of the last abbot, John Wakeman, who was consecrated first Bishop of Gloucester in 1541.   After the battle of Tewkesbury the Duke of Somerset and other Lancastrians took sanctuary in the church, but after two days they were dragged out and executed.

Deerhurst, the oldest church in the diocese, and one of the oldest in England, was the church of the Priory of that name, suppressed before the Dissolution.   The church was consecrated in 1056, and is of pre-Norman character, with a few later additions.   At the west end is a lofty tower, with small round-headed windows on the exterior.   Standing in the nave and looking back at the tower one may notice high up in its eastern face a very curious window of two lights, with triangular heads, each side of the head being formed of a single stone, very characteristic of this early style.   The chancel is entered from the nave by a small round-headed arch, such as may be seen at Bradford-on-Avon : the east end was a semi-circular apse, but the arch which led into it is now built up to form the present east end. Not far from the church is a Saxon chapel, called Odda's, which, like the Saxon church at Bradford-on-Avon, formed part of a dwelling-house, but was detected and restored to its proper state in 1884.

Deerhurst Church, Saxon Windows in Belfry

In thinly populated parishes the small Norman church, consisting only of nave, chancel, and bell-turret, often remains, as at Upper Swell, Condicote, Postlip, and the ruined Lancaut. In larger churches it is by no means

Odda's Chapel, Deerhurst, looking East

uncommon east of the Severn to find that the Norman doorways have been left, as at Elkstone, Lower Guiting, Withington, and Quenington.   The Early English style, with its beautiful lancet windows, and stiff-foliaged capitals, may be found, among other places, at Lydney, Almondsbury, Berkeley, and in the beautiful chancel at Lower Guiting.   A specimen of Decorated work will be

St Mary Redcliffe, Bristol

found in the choir at Tewkesbury, but both Early English and Decorated churches have generally been altered in the Perpendicular period by lowering the roof of the nave to build a clerestorey.

It is to the Perpendicular period that the finest churches of the county, apart from the abbey churches, belong—Chipping Campden, Northleach, Fairford, Painswick and Cirencester, to which (though in Somerset) we may add the great church of St Mary Redcliffe at Bristol. All were built by the great merchants, who in the later middle ages made such large fortunes out of wool, and their effigies may be seen cut in brass on the floor of the church, with their woolpacks, their shears and their sheep. All these churches have a family resemblance, though each

Cirencester Church, the South Porch

has its peculiar features.   Campden and Northleach agree in the arcades of the nave, which consist of lofty octagonal fluted columns supporting very depressed arches ; Fairford is famous for its wonderful glass, which has survived the devastations both of the Reformation and of the Civil War. The wool-merchant who built this church was John Tame, and he had the glass made to fit his windows, whether in England or in Flanders is a matter of dispute. Painswick and St Mary Redcliffe have their spires, but the latter—the child of the trade with foreign parts, rather than of homebred wool—stands alone for magnificence and splendour, and must not be classed with the humble churches of the country town.   Lastly Cirencester has its noble three-storied porch, and (like Burford in the neighbouring county) its many aisles.   This is not the place to speak of the havoc wrought in so many of our churches by the nineteenth century restorer.   There are signs that this false enthusiasm is now abating, and that a wiser generation will now treat our sacred monuments with the respect which they deserve.

18. Architecture—(*b*) Military. Castles.

The English before the Conquest did not build castles, nor did the Normans for some years after their arrival, except at one or two very important places, such as London. They contented themselves with fortifying a natural mound or constructing one for the purpose themselves. The summit of this mound they surrounded with a stockade, inside which was a kind of wooden building which served as a shelter and as a look-out.   In front of the mound was an enclosed court or "bailey," in which were the stables and the quarters of the soldiers.   But they were not long satisfied with such an elementary style of fortification; and soon began to build castles of stone. The strongest part of these stone castles, into which the garrison could retire when the rest of the defences were taken, was called the keep. This was most commonly made by surrounding the mound with stone, thus forming a kind of shell. But where the site was a new one, or sometimes where the mound was a natural one and therefore strong enough to bear the weight, the keep took the form of a massive rectangular tower, of which we may

Berkeley Castle

see examples at London, Rochester, and Norwich. If the mound was an artificial one and the shell keep was not adopted, the heavy rectangular keep would be built in the bailey.

In Gloucestershire only one of these keeps remains, the splendid shell keep of Berkeley. Gloucester and Bristol had rectangular keeps, but these are now destroyed.

There was once a Norman castle at Sudeley, but it has long ago disappeared, and the most ancient part of the present castle is no older than the reign of Henry VI, so that the only Norman castle remaining in the county is Berkeley. This castle has the distinction of having been inhabited by the same family from its foundation to the present day.

Some remains of the Edwardian or concentric type of castle may be seen at St Briavels and at Beverstone. The former was once a Norman fortress, but the earlier portions have long ago been destroyed, and the most conspicuous feature of the existing building is a gate-house with a semi-circular "drum" tower on either side of the gateway. It was always the residence of the Constable of the Forest, and is still Crown property. Beverstone was built by the Berkeleys in the fourteenth century on the site of an earlier castle. In the concentric type of castle the keep was no longer needed, and the fortifications consisted of an outer and inner ward, each defended by curtain walls and mural towers. A capital example of this type would be seen at Beaumaris, if the whole ruin were not smothered in a great mass of ivy, so that it is impossible to distinguish the distinctive features. Where the ground was rocky and did not accommodate itself to the strictly concentric type, the wards or courts were, so to speak, telescoped out, so that they came to be one behind the other, as at Chepstow. Where the castle was very large and powerful it was sometimes further strengthened by the addition of a third ward outside the other two, as at Caerphilly.

Here we must stop, but those who wish to know something about castles in connexion with the methods of attack and defence in use before the days of gunpowder will find it a most interesting study.

19. Architecture—(c) Domestic. Famous Seats, Manor Houses.

While the military architecture of the county might fill a volume the size of this, it would take fifty such volumes to describe its domestic architecture, so that in this chapter we can only give the merest outline of the subject.

The character of the houses in every district depends upon the kind of building material most easily procurable; thus in one place it will be stone, in another brick, in another a combination of brick and timber.   In Glouces-tershire, except in the northern parts of the Vale, it is stone.   East of the Severn it is the Oolite or yellowy grey stone of the Cotswolds; west of the river it is the Carboniferous Limestone of the Forest.   In both districts brick is an unwelcome intruder.   The most beautiful houses are of course in the former district, but those in the latter have a beauty of their own.   Up to a century ago, the limestone was not left bare as is now the case, but was covered with a thick coating of rough-cast and coloured white or yellow, the roof being high and gabled, and covered with burnt tiles or thatch, except in the few cases in which stone tiles could be afforded.   The rough-cast and wash kept the house dry and warm, for the naked rubble stone, though often seen nowadays, is not calculated to keep out either the damp or the cold. The whole harmonised perfectly with the landscape, and enhanced its beauty in the same degree that the modern style of house disfigures it.

The Market Hall, Minchinhampton

On the east side of the Severn the case is different. The houses of all sizes are built of the durable Cotswold stone—a far finer material—and most of them, except in the larger towns, date from the seventeenth century. The many fine manor houses of the Cotswolds, now generally used as farm-houses, are known far and wide, and the larger houses, such as those at Stanway and Bibury, are no less famous. Even the humblest cottages were built by the local mason who took pride in his craft, and put his own ideas into it; and there being no railways to convey the materials they were not "run up" by a builder with the ready-made productions of a distant factory. This Cotswold stone could easily be faced with the chisel, and soon became so hard and dry that it needed no plastering

The New Inn, Gloucester

or washing like the rough stone across the water. In a few years it assumed the delightful grey-brown tone which is the mark of every house from Campden to Bath, and from Banbury to Cheltenham. The tiles are shaped out of thin layers of sandstone which are quarried here and there in the Cotswolds, and also at Stonesfield in Oxfordshire. They are very heavy, and require a high pitched roof; a low roof could not bear their weight. In each tile a hole is bored to receive a wooden peg.

Of the Cotswold towns Chipping Campden, Stow-on-the-Wold, Northleach, and Cirencester are the most characteristic. The high street of Campden in particular consists from end to end of houses built in the fifteenth and sixteenth

Arlington Row

centuries, or even earlier—"Grevel's House" for instance is a splendid example of a fourteenth century merchant's house. In the older parts of Bristol may still be seen the tall half-timbered houses with the projecting upper stories, which were formerly the usual type of house in walled cities such as London and Exeter, and many others. Of late sixteenth or seventeenth century manor houses num-

bers are scattered throughout the district.   Earlier houses
are rare.   Southam de la Bere near Cheltenham was
built in the time of Henry VII; Thornbury and Down
Ampney in the time of Henry VIII; while Icomb Place
on the eastern declivity of the hills and Wanswell Court
near Berkeley belong to the fifteenth century.   The
former has been altered and restored, but the latter, which

Owlpen Manor House

may be taken as a type of the smaller manor house of the
period, apart from a wing added in the reign of Elizabeth,
remains as it was built.   The house faces south and a
south porch with a plain pointed arch and a room over
it opens directly into the large square hall, which served
as a dining-room for the household.   In the north-east
corner a door opens into a parlour with a good perpen-

dicular window of four lights to the south and a slanting
opening on either side of it, one of which commands the
front entrance, and the other the approaches in the other
direction.   The kitchens and cellar or larder are behind
the hall to the north, with the bedrooms over them.
Another small parlour with a bay window is on the west
side of the hall, while the Elizabethan oblong building is
attached to the western side increasing the frontage only
by its width.   The original house was probably built in
the middle of the fifteenth century about the time of the
close of the French war.   The time for the building of
fortified houses came to an end with the invention of
gunpowder, and half a century later the castles of the
feudal nobility commonly fell into decay.

20. Communications—Past and Present. Roads, Railways, Canals.

The earliest roads would be the clearings made by the
earliest inhabitants through the woods which then covered
the hills.   Of these of course we have no knowledge, but
the British trackways of later times may have often
followed the same lines.   Of these trackways many traces
are still left, for instance in the hollow ways so common
in the west of the county.   Sometimes however these
hollow ways merely mark the track of the old road before
the days of Macadam, and may now be seen in some
places following the course of the new road at a lower
level : these may, or may not, date from pre-Saxon times.

When the Romans came, they sometimes built their great military roads along the course of the old trackways, sometimes they struck out a new line for themselves. Some of these roads are still great highways of our

The Fosseway near Stow

county. The chief is the Fosseway, which connected Exeter with Lincoln : it runs in an almost straight line over the hills from Cirencester to Moreton. Another is Ermine Street, which comes from Newbury to Cricklade in Wiltshire and continues thence as the main road through Cirencester to Gloucester. From Birdlip Hill you may see it below you like a long white line, straight from the foot of the hill to the city. Then there is Akeman Street, which connects Colchester with Bath: it passes through St Albans and Alcester near Bicester, enters the county at the bridge over the Leach, and joins the Fosse at Cirencester. In this county, except for three miles about Coln St Aldwyn, it coincides with a main road. The Via Julia from Bath to Caerwent crossed the Severn somewhere above the site of the present tunnel. There are also many cross roads, but the only one we can mention here is the road which joins Gloucester and Caerwent, crossing the Wye at Chepstow about half a mile above the present bridge, where the remains of blackened stakes are to be seen at low water of spring tides.

From the Conquest onwards, when the monasteries took charge of the high roads, it is probable that they were in tolerable order, at least for horse traffic, but when once the monks and feudal labour had departed their state became deplorable. This is what a competent observer in the latter half of the eighteenth century says about the roads of the Vale: they "are shamefully kept. The Parish roads [those kept up by the parish] mostly lie in their natural flat state with the ditches on either side of them full to the brim. The toll-roads are raised (generally much too high) but even on the sides of these I have seen full ditches....The foundation is ever a quagmire; and the superstructure, if not made unnecessarily strong, is always liable to be pressed into it. Hence the deep ditch-like ruts which are commonly seen in roads of this description. The road between Glocester and Cheltenham (now become one of the most public roads in the island) is scarcely fit for the meanest of their Majesties' subjects to travel on....and pay for; much less suitable for their Majesties themselves, and their amiable family, to trust their own persons on." The "toll-roads" here spoken of were the roads on which a toll was collected at gates called turnpikes which barred the road : the gates have long been gone, but the toll-houses occasionally remain. The gates of a fixed district were let to the highest bidder by a "Turnpike Trust," which also kept the roads in repair. These roads as well as the parish roads are now managed by the County Council.

The high roads of the hills have wide grass margins which were formerly used as the summer road, thus keeping the metalled road free from wear and tear till the winter rains set in, but the writer just quoted complains that the practice had been adopted of digging pits in these grass margins, out of which the stones for the repair of the road were procured. "The materials" however, he says, "are more plentiful than durable : presently grinding down under heavy carriage." Another writer at the beginning of the nineteenth century finds the roads much improved (these were the days of Macadam) and reports that in the neighbourhood of Gloucester the limestone from the rocks at Clifton and Chepstow had already come into use for metalling. At the present day the high roads throughout the county are metalled with a hard stone from Clee Hill in Shropshire and are some of the best in England.

E. G.   8

At the beginning of the last century the great coach roads were the London, Oxford, and Cheltenham road, which enters the county near Burford, and passes through Northleach to Cheltenham, and thence on to Gloucester: the Oxford, Chipping Norton, and Worcester road, which cuts across the northern extremity of the county from

Old Bridge over the Avon at Tewkesbury

Moreton through Bourton-on-the-Hill to Broadway, above which place it descends the steep hill by a zigzag, while the course of the older and straighter road may be seen hard by : the Faringdon, Lechlade, Cirencester, Tetbury, and Bath road, which keeps to the flat country on the eastern slopes of the Cotswolds, and enters Somerset about five miles north of Bath : the Birmingham, Tewkesbury, Gloucester, and Bristol road, which keeps to the foot of the western escarpment of the Cotswolds: the Hereford, Ledbury, and Gloucester road: the Brecon, Abergavenny, Monmouth, Ross, and Gloucester road: and lastly the Bath, Bristol, Newport, and South Wales road, which crossed the Severn by a ferry either at the New Passage at Redwick, or three miles higher up by the Old Passage at Aust. There were of course many other roads traversed by coaches, but those mentioned were the chief mail routes.

By 1840 the railway had found its way into Gloucestershire. The first lines to be constructed were the Great Western from Swindon to Gloucester, and the Midland from Birmingham to Gloucester and Bristol. The former enters the county near Thames Head, penetrates the Cotswolds by the Saperton tunnel, and runs down the Stroud valley to Stonehouse, whence it continues parallel with the Midland to Gloucester. The latter crosses the Avon at Defford, enters the county at Ashchurch and runs past Cheltenham to Gloucester, and thence by Wickwar and Yate to Bristol. Then came the G.W.R. line from Oxford to Worcester by Adlestrop, Moreton, and Campden; the line from Gloucester to Grange Court, Ross and Hereford; and the South Wales line to Lydney and Chepstow. Later were made the Great Western branch lines from Kemble to Cirencester on the one side, and to Tetbury on the other; the Wye Valley railway from Chepstow to Monmouth which is in Gloucestershire as far as Tintern, where it crosses the Wye; the line from Gloucester to Ledbury by

8—2

Newent and Dymock; and lastly (though not in time)
a line through the Cotswolds (till then untouched by any
railway) from Cheltenham to Bourton-on-the-Water,
which was subsequently continued through Chipping
Norton to Banbury. Many years after this railway was
opened, the Swindon and Andover railway was continued
northwards from Swindon, through Cirencester and
Withington to Andoversford, from which place the
trains run over the G.W. line into the Midland Station
at Cheltenham. The line was opened as the Midland
and South Western Junction Railway. It enters Gloucestershire just after passing Cricklade.

The Severn and Wye railway from Berkeley through
the centre of the Forest to Lydbrook crosses the Severn at
Sharpness by a light high-level bridge. From Coleford
there is a branch of the Wye Valley railway to Monmouth
through Newland. The Cheltenham-Honeybourne branch
now forms part of the G.W.R. express route from Cornwall
to the north-west of England. There is also a light railway from Moreton to Shipston-on-Stour, part of which
passes through the north-eastern projection of our county.
The G.W. branch from Bristol to South Wales formerly
crossed the water by a steam ferry, but in July 1887 the
famous Severn Tunnel took its place. This great tunnel,
4 miles 624 yards long, of which 2½ miles are under water
at high tide, took nearly 14 years to make, the work having
been more than once brought to a standstill by the tapping
of fresh water springs; but the story of the tunnel would
require a whole book much larger than this one. To
make the list of the Great Western railways of the

county complete it should be added that the four last miles of the Oxford, Witney, and Fairford branch are in Gloucestershire, that lines from Bristol and Clifton run along the right bank of the Avon to the Docks at Avonmouth, and that a new South Wales route has been opened from Swindon *viâ* Badminton and the Severn Tunnel.

The Midland branches are—from Ashchurch on the west to Tewkesbury and Malvern, and on the east to Evesham. Detached portions of both these railways are within our county. Other branches run from Stonehouse to Stroud and to Nailsworth ; from Coaley Junction to Dursley; from Berkeley Road to Berkeley; from Yate to Thornbury, and from Mangotsfield to Bath.

It should be added that iron railways for the conveyance of trucks of coal or iron drawn by horses had been in use in the Forest at least as early as the beginning of the last century.

Canals began to be constructed in the last quarter of the eighteenth century, but though they brought manure, lime, and other necessaries, they were at first objected to by farmers as diminishing the quantity of land available for cultivation, and a writer in 1813 calculated that in Gloucestershire they had occasioned the loss of 515 acres. In spite of this they were a great advantage to the community in general for the conveyance of coal and all other heavy goods, the freight being much less than that charged for land transit. The eager promoters of the canals little thought that in another half century they would be superseded by a faster and easier means of conveyance.

Nevertheless the canals remained serviceable for heavy goods such as coal when speed was not desired, and in the Midlands the barges still have a considerable traffic, especially in coal and hay. There are even signs, as we shall see directly, of the revival of the canal system on the ground that the carriage of heavy goods is cheaper by water than by rail.

The principal canal of the county is the great Gloucester and Berkeley ship canal. Until the formation of the Manchester ship canal it was the biggest in England. It was considered a colossal undertaking when it was first started, and although begun in 1794, a writer of 1813 reports that though £120,000 had been expended scarcely 17½ miles were finished. Finished however it was at last in 1827. It is 18 ft. deep, and can carry ships of 1200 tons up to Gloucester. The distance from Sharpness to Gloucester is 16 miles by the canal, and 28½ by river. In 1875 the new docks were built at Sharpness, in which ships of 5400 tons can be docked.

The idea, long entertained, of uniting the Thames and Severn was realised at last by the Thames and Severn Canal, completed in 1789, when the first boat laden with goods passed into the Thames. The canal is nearly 29 miles long. It begins at Framilode and passes up the Stroud valley to Saperton tunnel, the entrance of which is a mile eastward from the entrance to the railway tunnel: after emerging from the tunnel, which is nearly 2½ miles long, it drops down into the Thames Valley and joins that river at Lechlade. Its traffic in 1841 when the G.W.R. was opened from Swindon to Cirencester was

89,271 tons. It was then bought up by the railway company, who closed it in 1893. By this time the tunnel was much out of repair, but at last it was taken up by a public trust, and handed over to the County Council in 1901, and it is now open between Framilode and Saperton.

Mouth of the Canal Tunnel, Saperton

The Hereford and Gloucester canal (1792–1798) is now derelict and is largely absorbed by the Gloucester and Ledbury railway. It has been suggested that, if the Worcester and Birmingham Canal were enlarged and deepened, large vessels might by means of its connexion with the ship canal be taken up to Birmingham, thus opening the Atlantic to the centre of England. All such questions as this have recently been the subject of a Royal Commission on Canals and Waterways who issued their Report in 1907.

The Thames and Severn Canal. At the Blue House
above Siddington

21. Administration and Divisions.

When a strange race settle in a new country, they naturally bring with them their social and political institutions. This was the case with the English when they settled in America in the sixteenth and seventeenth centuries, and with the Saxons when they settled in Britain in the fifth and sixth. Henceforth the land was called

England, and the people English. The earliest political division of the English was the Hundred and the earliest unit the Township. The Hundred, according to some authorities, was the district, larger or smaller, occupied by a hundred fighting men with their families. Where the population was scanty the Hundred would be larger, where dense, smaller. The township had a chief officer called the Reeve, and settled small disputes for itself. Larger matters were taken before the Hundred Court, which met once a month and was attended by the Reeve and four freemen from each township and also by the Eorls or Thegns residing in the Hundred. Above the Hundred Court was the court of the whole folk or tribe, consisting of several Hundreds. This Folk Court met twice a year and consisted of the members of the Hundred Courts, presided over by the Ealdorman. When the parishes were instituted by Archbishop Theodore in the latter part of the seventh century they were made to coincide with the townships. The name township is seldom used nowadays, except in legal phraseology, but where its ecclesiastical organisation has been preserved, it is simply the civil name for the parish.

Later, when the shires or counties caused the combining of several Hundreds, the importance of the Hundred Court declined, and that of the Folk Court—which is now the Shire or County Court—increased. By the time the Midland shires were formed the town from which each took its name had become a place of importance, and the Folk Courts were probably held there.

It must be noted that since the Domesday Book was

compiled, some of the Hundreds have changed their names.
Originally the name was either that of some great chief
or was taken from the place—a great tree for instance,
or a riverside—where the Hundred Court was held, but
the name has often survived in some modern village. It
would be an interesting task to take the Hundreds of
some particular county and attempt to trace their fortunes
from the earliest times.   Here is a list of some of the
present Gloucestershire Hundreds, the name of which is
not at once apparent :—

Bledisloe	containing	Lydney and Awre	The word *loe* is a variant of low, here probably meaning a burial mound
Botloe	„	Newent	
Barton Regis	„	Mangotsfield	The King's demesne
Bradley	„	Northleach	Broad lea
Brightwell's Barrow	„	Fairford	Like loe, a burial mound
Crowthorne	„	Cirencester	
Dud-Stone and King's Barton	„	[Gloucester]	The watch stone
Grumbald's Ash	„	Badminton	A tree
Kiftsgate	„	Campden	A road or an opening
Langley and Swineshed	„	Almondesbury	
Longtree	„	Minchinhampton and Tetbury	
Tibaldstone	„	Ashton-under-Hill and Beckford	
Whitstone	„	Fretherne and Stonehouse	

The above list shows that trees, stones, and burial-
mounds were some of the commonest meeting-places for
the Hundred Court in the earliest times. The number of

Gloucestershire Hundreds is now 28, a large number compared with some of the adjacent counties. Thus, though Wiltshire has 29, Worcestershire and Warwick have only five each, Oxfordshire 14, and Herefordshire 11. The jurisdiction of the Hundred has long been obsolete, but it still has a constable or bailiff, and in case of damage wilfully done by rioters, the owner of the property can sue the Hundred for damage.

Gloucester and Bristol are counties of themselves and have their own assizes, but, generally speaking, the old County Courts are now represented by the Quarter Sessions and the County Council.

The Quarter Sessions are held by all the magistrates of the county that choose to attend ; they meet at Gloucester. Their functions are now almost reduced to criminal business and licensing public-houses. Their other functions, such as the management of roads and bridges and the control of Lunatic Asylums, have since 1888 passed to the County Council, which also meets at Gloucester. The police are controlled by a joint standing committee of the Quarter Sessions and the County Council. The County Council consists of 75 members, of whom 19 are aldermen, who during their term of office do not require re-election ; this gives an element of permanence to the Council. The modern County Courts were instituted in 1846 for the settlement of small debts : these County Court districts are mapped out irrespective of county boundaries.

Below the County Council are the Urban District Councils and the Rural District Councils. As an example of the latter we will take the Lydney Rural District

Council, which comprises the parishes of Lydney, Alvington, St Briavels, Hewelsfield, Woolaston, Tidenham, Lancaut, and the "tithing" of Aylburton : the population of this district is 8580.

Below the District Councils are the Parish Councils. The ancient parish or township had always formed a unit for administrative purposes, but now the civil parish[1], which may or may not coincide with the ecclesiastical parish, is the unit. There are 359 civil parishes in the county; the ecclesiastical parishes are more numerous.

There are 15 Unions for the administration of the Poor Law, and some border parishes belong to the Union of the adjacent county.

The municipal boroughs are Bristol, the population of which in 1921 was 377,061 : Cheltenham, 48,444 : Gloucester, 51,330 : Tewkesbury, 4704. All, except Cheltenham, have a separate jurisdiction. The Hundred of the Duchy of Lancaster, containing Mitcheldean and Minsterworth, is for some purposes within the jurisdiction of the Duchy.

We have left to the last the two chief officers of the county and its Parliamentary representation. The two chief officers are the Lord Lieutenant and the Sheriff, i.e. the Shire-reeve. The one may be said to represent the king and the other the county. Nevertheless both have for many centuries been appointed by the Crown. The Lord

---

[1] A civil parish is the area for which a separate poor-rate is, or can be made, or for which a separate overseer can be appointed. Extra parochial places were abolished by the Statutes of 1857 and 1868.

Lieutenant is appointed for life, and is also the Custos Rotulorum: he is the head of the magistracy and recommends to the king persons fit for appointment as justices. Formerly he had the control of the militia, but after 1871 this was transferred to the Crown. The Sheriff is the executive officer of the county; he is chosen from the wealthy commoners every year on the morrow of St Martin's, November 12th. It is his duty to attend the judge at the assizes, and see that his sentences on the prisoners are carried into execution. Neither he nor the Lord Lieutenant is a paid officer. It is the Sheriff, too, who presides over the election of members of Parliament, and who applies to the Speaker of the House of Commons for a writ in the case of a vacancy.

Gloucestershire returns 11 representatives to the House of Commons—seven borough members and four county. The boroughs are—Bristol five, Gloucester one, and Cheltenham one. There are four county divisions, each returning one representative, viz. the Forest or Newnham, Cirencester and Tewkesbury, Mid or Stroud, and South or Thornbury.

The diocese of Gloucester comprises the whole county, except Bristol and four parishes besides, which are in the diocese of Bristol. There are two archdeaconries, Gloucester and Cirencester, and each archdeaconry is divided into rural deaneries. The archdeacon has a general jurisdiction over his district under the bishop. It is the business of the rural dean to call together at stated periods the clergy of his deanery, and to see that its churches and their furniture are in good order.

22.   Roll of Honour of the County.

Gloucestershire can claim an unusual number of famous men, at the names of the more noteworthy of whom we may glance in this chapter.

Let us begin with the divines.   Except for 63 years

Joseph Butler

in the nineteenth century Gloucester and Bristol, two of the sees created by Henry VIII in 1541, have always been separate dioceses.   Among the bishops of Gloucester may be mentioned John Hooper (bishop 1550–54), who, a martyr to the Protestant faith, was burnt at Gloucester in 1555 ; and William Warburton (1759–79), the author of the *Divine Legation of Moses*, and the friend and executor of Pope. Of the bishops of Bristol we will only mention Sir Jonathan Trelawny (1685–89). He was one of the

John Keble

seven bishops, who with Archbishop Sancroft at their head, petitioned against James II's Declaration of Indulgence, and is the hero of Hawker's ballad *And shall Trelawny die?* Joseph Butler (1738–50), one of the best known of English divines as the author of the *Analogy*,

was bishop from 1738 to 1750 and was afterwards trans-
lated to Durham. The great churchman, John Keble,
was born at Fairford in 1792. He was one of the leaders
of the "Oxford movement" of the thirties, and was
professor of Poetry at Oxford from 1831–42. He is
now best known as the writer of *The Christian Year*, first
published in 1827. He died in 1866.

As Gloucestershire statesmen or lawyers may be mentioned Sir Matthew Hale (1609–76), Lord Chief Justice
of England, who was born at Alderley, educated at
Magdalen College, Oxford, studied the law at Lincoln's
Inn, represented his county in Parliament, and in 1671
was made Chief Justice of the King's Bench : Henry,
second Earl Bathurst (1714–94), was Lord Chancellor
1771, and son of the famous first earl, the friend of Pope.

The historians have been few. Robert of Gloucester
wrote a rhyming Chronicle of Britain in the time of
Edward I, and Richard of Cirencester (died 1401), so
called from the place of his birth, was a monk of Westminster, where he wrote the *Historia ab Hengista*. The
Itinerary ascribed to him, however, is spurious. The
brothers Daniel (1762–1834) and Samuel (1763–1819)
Lysons planned a history of the counties of England in
alphabetical order, but they only got as far as Devon-
shire. Daniel, who inherited Hempstead Court on the
death of his uncle, became F.R.S. and F.S.A. In 1811
he published a valuable work entitled *The Environs of
London*. Samuel was keeper of the archives in the Tower
of London and F.R.S. He published, 1801–17, *Reliquiae
Britannico-Romanae*, and in 1803, a folio volume of en-

gravings entitled *A Collection of Gloucestershire Antiquities.*
Of county historians we have Sir Robert Atkyns of Saperton
(1647–1711). His now scarce folio was issued the year
after his death. Samuel Rudder (1726–1801), whose
folio was published in 1779, improved upon Atkyns, and
brought his history up to date. He also wrote a history
of Cirencester. Thomas Rudge, B.D. (1754–1825), was
the son of Thomas Rudge of Gloucester : he was educated
at Merton and Worcester Colleges, Oxford, and became
archdeacon of Gloucester in 1814 : he wrote a compressed
History of the County of Gloucester and a Survey of the
agriculture of the county. Thomas Dudley Fosbroke
(1770–1842) was another of the distinguished members
of Pembroke College, Oxford. He was curate (1810),
and vicar (1830), of Walford near Ross, and wrote a
History of the City of Gloucester and other archaeological
works. Samuel Lysons the younger, the son of Daniel
already mentioned (1806–77), wrote a work on *The Romans
in Gloucestershire* which was published in 1860.

A famous merchant was William Grevel of Chipping
Campden (died 1401). He was a prince among wool
merchants and his house at Campden is still the finest in the
town. Great Bristol merchants were William Canynges
(1376–96) and his grandson William Canynges the
younger, five times mayor. The latter died in 1474,
a priest in the college of Westbury-on-Trym. Edward
Colston (1636–1721), founder of the Free School, is
perhaps one of the best known of past Bristol worthies;
his anniversary day falls on the 13th of November, and
is celebrated by a banquet. Sebastian Cabot (b. 1474 ?)

was probably a native of Bristol : his father, John Cabot, a Genoese, had settled there ; they were both great navigators and explorers, and to them we owe the discovery of Labrador.   Sir William Penn (1621–70), father of William Penn, the founder of Pennsylvania, was born at Bristol : he rose to the rank of admiral, distinguished himself in the Dutch wars, and captured the island of Jamaica, which has ever since remained a British possession.   John Taylor, the " water poet " (1580–1653), was born at Gloucester, and began life as apprentice to a London waterman.   His verses excel in number rather than merit, and are valuable rather on antiquarian than poetical grounds.   Thomas Chatterton (1752–1770), " the immortal boy," was the son of a chanter of Bristol Cathedral ; in the Treasury of this church he pretended to have found the MSS. which he afterwards printed as the composition of a monk of the fourteenth century called Rowley.   With brilliant abilities, he might have won success in a more legitimate field, but poverty and misfortune drove him to suicide at the age of 18 in a London garret.   Robert Southey (1774-1843) was the son of a linen-draper in Bristol : he was educated at Westminster and Balliol College, Oxford.   In after life he settled at Keswick and was the staunch friend of Wordsworth and Coleridge.   He is the best known Gloucestershire poet, though he is a better master of prose than verse.

Gloucestershire has produced more scientists than poets. Edward Jenner (1749–1823), the discoverer of vaccination and pioneer of modern sero-therapy, was the son of the vicar of Berkeley. He was educated privately, and afterwards took the degree of Doctor of Medicine and was elected a Fellow of the Royal Society. In his latter years he resided at Berkeley, where he died. James Bradley, D.D., F.R.S. (1693–1762), was born at Sherborne in the

Edward Jenner

Cotswolds, and educated at Northleach Grammar School and Balliol College, Oxford. He became Savilian Professor of Astronomy in the University of Oxford, and in 1742 Astronomer Royal. John Lightfoot (1735–1788) was born at Newent, educated at the Crypt School,

9—2

Gloucester, and at Pembroke College, Oxford. He took Holy Orders, and became a Fellow of the Royal Society. His great work was the *Flora Scotica*, published in 1788, and he was the first to distinguish and record as a British bird the reed warbler, which he observed on the Coln near Uxbridge. John Caxton, F.R.S., the electrician (1718–72), the first to make powerful magnets, was born at Stroud. Jonathan Hulls, the inventor of the steamboat, was the son of a mechanic at Aston Magna, a hamlet near Moreton. He settled at Broad Campden as a clock repairer, but after many years of hard work, and partial success, he retired, for want of money and encouragement, to London, where he died.

William Tindale, the translator of the New Testament, was a member of a Gloucestershire family. The branch to which he belonged was settled at Stinchcombe. He was born about 1495 and after a strenuous life was executed as a heretic at Vilvord in Brabant in 1536. A monument has been erected to him on Stinchcombe Hill.

George Whitefield (1714–70), the Calvinistic preacher, was born at the Bell Inn in Gloucester : he was educated at the Crypt School and Pembroke College, Oxford. His addresses to the colliers at Kingswood had a great effect upon this, at that time, wild and ignorant people. He was a clergyman of the Church of England, and made more than one visit to the American colonies, where he died.

Robert Raikes (1735–1811), the promoter of Sunday Schools, was born at Gloucester. He was the son of Robert Raikes, the founder, in 1722, of *The Gloucester*

Journal, one of the earliest local newspapers; on his father's death in 1757, the son succeeded to the business, and as a newspaper editor and an energetic doer of good works he spent the rest of his days in his native city.

Sir Thomas Lawrence

Lastly we come to men of letters or art. William Grocyn (1442–1519), the tutor of Sir Thomas More and of Erasmus, and first teacher of Greek at Oxford, was born at Bristol, and educated at Winchester and New College, Oxford. Greek was almost unknown in England when Grocyn went to Italy in his 46th year to perfect himself in the language.

Richard Graves (1715–1804) was the son of Richard Graves, the antiquary of Mickleton. He was educated at Abingdon School and Pembroke College, Oxford : he was afterwards elected to a fellowship at All Souls. In 1750 he became rector of Claverton near Bath, where he resided till his death. He is the author of *The Spiritual Quixote* and other amusing tales. He was the friend of the poet Shenstone, of whom he wrote his *Recollections.*

Hannah More (1745–1833) was born at Stapleton. She began life as a writer for the stage, and made the acquaintance of all the principal literary men of the day. At the age of 41 she retired to Barley Wood, a house near Wrington in Somerset, and devoted herself to writing moral and religious books.

In the domain of art our county takes a very high place. Bristol can claim a very long list of artists of note, and though no "School" was formed as at Norwich, their work was scarcely less remarkable. Nicholas Pocock (1741–1821), may be reckoned as one of the founders of English water-colour painting, though mainly a marine artist, and Edwin Hayes (1820–1904) is even better known for his sea pieces. George and Alfred Fripp, relatives of Pocock, also water-colourists, won considerable fame, as did also James Baker Pyne (1800–1870). The greatest artist, however, that Bristol produced was William James Müller (1812–1845) who attained the highest rank as a painter both in oil and water-colour. Sir Thomas Lawrence the famous portrait painter (1769–1830) became President of the Royal Academy in 1820.

23. THE CHIEF TOWNS AND VILLAGES OF GLOUCESTERSHIRE.

(The 1921 population figures are given for all towns for which they are available, but for many of the smaller places the 1921 figures have not yet been published.)

Almondsbury (2213), a popular and increasing village, eight miles from Bristol. On the slope of the hill is Knole Park, for 300 years the property of the Chester family. (pp. 100, 122.)

Amberley, an ecclesiastical parish, formed in 1840 out of the parishes of Minchinhampton and Rodborough. It is a health resort at an elevation of about 500 feet one mile west from Minchinhampton. At Rose Cottage, Miss Mulock, afterwards Mrs Craik, wrote her famous novel *John Halifax, Gentleman.*

Awre with **Blakeney** (1147). Awre stands on a peninsula of the Severn opposite Fretherne. Blakeney, separated as an ecclesiastical parish in 1877, is famous for its orchards and is one of the six woodwardships of the Forest. (p. 20.)

Barnwood (1337) is now a suburb of Gloucester, with a large hospital for the insane.

Berkeley (826), famous for its Norman castle, which has been inhabited by the same family from the twelfth century to the present day. Edward II was murdered within its walls. The thirteenth century church has a detached bell-tower. Dr Jenner, the discoverer of vaccination, was born here. Sharpness, a modern settlement, at the entrance of the Gloucester and Berkeley canal, is also the eastern extremity of the high-level Severn railway bridge. (pp. 11, 20, 28, 93, 100, 103, 117, 118, 133.)

Bishop's Cleeve (657), three and a half miles from Cheltenham. Cleeve Hill (1070 feet) in this parish is the highest point of the Cotswolds. The Cotswold convalescent home on the summit was erected by public subscription in 1903–4. There is a fine Norman church and an Elizabethan rectory house.

Bisley with **Lypiatt** (1936) is an ancient parish at a height of 790 feet, four miles from Stroud, which was separated from it in the reign of Edward II.

Bitton (3244), halfway between Bristol and Bath, is a large parish on the Avon. The pretty village has a fine church with a tower built in 1337. There are also coal-mines of great depth. (pp. 6, 22.)

Bourton-on-the-Water (1153), a pretty village on the Fosseway between Northleach and Stow-on-the-Wold. The Windrush flows down the main street with houses on either side of it. (p. 116.)

Bream, a small village—half mining, half agricultural—on the borders of the Forest, four miles from St Briavels. See Newland.

Brimscombe, an ecclesiastical parish formed in 1840 out of the parishes of Minchinhampton and Rodborough. The Great Western Railway has a station here.

Bristol (377,061) was the second city in England until the rise of the northern manufacturing towns, and is still the chief city in the West. Its great merchants, such as William Canynges and his grandson of the same name, and Edward Colston—names that are household words with every Bristol man—lavished their wealth for the good of their native city. Its trade with Spain and the West Indies was hardly surpassed by that of London; and a great impetus has lately been given to the West Indian trade owing to modern inventions, which make it possible to import fresh fruit in large quantities. Next to the merchants came the vintners, and Bristol sherry is still some of the finest that can be procured. Vessels of considerable size are able to come up into the very centre of the city, and large docks have recently been constructed at Avonmouth, which will hold still bigger ships. In Bristol, about 1490, settled John Cabot with his three sons, of whom Sebastian is best known. He made voyages from thence and was the first to sight the mainland of America. The finest church in Bristol is St Mary Redcliffe. The magnificent thirteenth century tower had formerly an unfinished spire, but this was completed in 1872 by the citizens. The main body of the church is Perpendicular, but there are some beautiful Decorated portions, notably the richly undercut exterior arch of the north porch. It was in the "Treasury" over this porch that Chatterton pretended to have found the MS. which he afterwards printed as the "Rowley" poems. Except from 1836–1897 Bristol has been the head of a diocese since 1542; the greater part of this diocese is now in Wiltshire. The cathedral, before the Reformation the church of a monastery of Austin Canons, consisted until 1877 of a choir and transepts only: in that year the modern nave was finished. The rest of the building is of various dates, almost the oldest being the Transition Norman Chapter-house. The Bishop's Palace was burnt by the Reform rioters of 1831. The magnificent "College gateway" with a splendid Norman arch is a survival of the monastic buildings destroyed at the Reformation. Bristol has a Lord Mayor, and returns five members to parliament. Clifton, once a village, is now the residential quarter of the city; it lies on the cliffs and downs to the north. The chain suspension bridge across the gorge of the Avon was begun in 1860, when the chains of the Hungerford Suspension Bridge in London were purchased for the purpose. By the river-side at the foot of the cliffs stood the Hot Wells, a building which covered a medicinal spring, formerly much resorted to for chest complaints, and here too was the landing place for the Chepstow steamers. Clifton College was founded in 1862. (pp. 4, 42, 43—52, 57, 66, 69, 71, 74, 89, 93, 101, 108, 113, 123, 124, 125, 127, 129, 130, 131, 132, 134, 135.)

Cainscross (2190), a modern parish constituted in 1894 out of the adjacent parishes. It is in the clothing district between Stroud and Stonehouse.

Clifton and the Suspension Bridge

Cam (1834), one mile from Dursley. Lower Cam, with a large cloth mill, was cut off as an ecclesiastical parish in 1888.

Chalford (2913), an ecclesiastical parish, with a manufactory of walking and umbrella sticks, formed out of Bisley in 1842. (p. 58.)

Charlton Kings (4361) is now a suburb of Cheltenham and a favourite place of residence. It has two churches; the modern one, opened in 1871, is remarkable for its numerous carvings in stone.

Cheltenham (48,444), formerly a small country town, but since the time of George III, whose first visit took place in 1788, a fashionable watering-place. Its mineral springs were first noticed in 1718. There are several Pump Rooms, including the Montpelier built in 1826 close to the Promenade, and the Pittville built at the same period by Joseph Pitt, M.P. Since 1894 the latter has been surrounded by Pittville Park. Cheltenham College dates from 1841, and the Ladies College from 1854. The town is a

Cheltenham College

borough and returns one member to Parliament. (pp. 10, 21, 57, 61, 109, 113, 114, 115, 116, 124, 125.)

Chipping Campden (1680), an ancient town in a hollow of the northern extremity of the Cotswolds, and, like Northleach, one of the chief centres of the wool-trade. Few of the houses of the long wide street are later than the seventeenth century, and several are still older. The magnificent fifteenth century church stands at the east end of the town hard by the ruins of Campden

House, burnt in the civil wars. The seventeenth century alms-houses built by the first Viscount Campden are remarkable for their strong and simple façade and their raised position above the street. (pp. 9, 19, 65, 67, 100, 101, 107, 108, 122, 129.)

Chipping Sodbury (977), a small town eleven miles from Bristol.

Churchdown (1126). A favourite residential place midway between Gloucester and Cheltenham. The church, on Chosen Hill, is a landmark for miles round.

Cirencester (Ciceter) (7408) with its fine wide market-place and stately church may well be called the capital of the Cotswolds. The Roman town is now buried under the modern one, and so thorough was the destruction of the former that even the streets do not correspond. Of the wealthy abbey, like Bristol a house of Austin Canons, hardly a trace remains. The church is mainly a fifteenth century edifice; the three-storied porch facing the market-place is its most striking feature. The famous Agricultural College was founded in 1845. Close upon the town are the gates of Oakley Park, the seat of Earl Bathurst, a large tract of woodland and rough pasture reaching to Saperton. (pp. 17, 67, 75, 76, 87, 88, 89, 102, 108, 111, 112, 114, 116, 118, 122, 125, 129.)

Clifton. See Bristol.

Coalpit Heath, in the Bristol coalfield, is an ecclesiastical parish formed in 1845 out of the parishes of Westerleigh and Frampton Cotterell.

Downend, in the Bristol coalfield, an ecclesiastical parish cut out of Mangotsfield in 1874.

Dursley (2601), a small town, four miles east of Berkeley, at the foot of Stinchcombe Hill. It has a branch railway from Coaley Junction. Agricultural implements are manufactured here. (p. 117.)

Dymock (1297), in the extremity of the Newent district, celebrated for its orchards. The Man of Ross was born here. (pp. 22, 116.)

Eastington (1202), one of the clothing villages one mile from Stonehouse. The fifteenth century church was enlarged in 1860. It must not be confused with Ampney St Peter, which is commonly called Eastington.

Fairford Church

Fairford (1410), on the Coln with a branch railway from Oxford, twenty-seven miles. In 1850 an Anglo-Saxon cemetery containing various relics was found near the town. The fine church was rebuilt by John Tame, wool-merchant, in the fifteenth century. It is famous for its windows of coloured glass. This wonderful glass, which fills twenty-eight windows, was made to

fit them—perhaps in Flanders. It only survived the devastation
of the Civil War by being taken out and buried by a patriotic
parishioner. Cirencester is eight miles west by road. (pp. 17,
32, 60, 100, 101, 122.)

Forest of Dean. This large forest was formerly extra-
parochial, but in the last century it was divided into ecclesiastical
parishes, viz. Christchurch, formed in 1844 out of the township
of West Dean, a suburb of Coleford: Coleford (see under Newland): Cinderford, with two parishes, one formed in 1844, the
other in 1880: East Dean (14,594), on the high road between
Mitcheldean and Newnham, containing three ecclesiastical parishes,
and West Dean (10,570) containing four: Viney Hill, formed out
of the townships of East and West Dean in 1866; it includes part
of the hamlet of Yorkley: Parkend, formed out of the parish of
St Paul in 1842. It is four miles from Coleford and contains
some of the best coalpits in the Forest. (pp. 5, 6, 13—15, 29,
30, 33, 36, 71—74, 88, 104, 116, 117.)

Frampton Cotterell (2068), in the Bristol coalfield seven
miles from Bristol. There is a hat manufactory in the parish.

Frenchay. A well-built village four miles from Bristol: it
was constituted an ecclesiastical parish out of Winterbourne in 1830.

Gloucester (51,330), an ancient city and the capital of the
county. The cathedral was formerly the church of St Peter's
Abbey, but was preserved as the cathedral church of the new
diocese, when the monasteries were destroyed. The central tower
with its perforated pinnacles and battlements is one of the most
beautiful in England. The whole church is Norman, with some
later additions. The imposing Norman arcade of the nave
resembles that of Tewkesbury, and is untouched, but over the
Norman work of the choir is thrown a delicate veil of Perpendicular tracery which conceals its original state. The windows
are also of the later date. The east window is second only to
that of Carlisle cathedral in size. The vaulting of the choir is

one of the richest in England. Among the tombs may be mentioned that of Edward II, whose murdered corpse was brought here from Berkeley. Gloucester is a borough, returns one member to Parliament, and is a county in itself. It is the chief port on the Severn, and has a considerable trade both inland and with foreign countries. (pp. 2, 3, 8, 11, 20, 39, 41, 42, 57, 61, 69, 74, 76, 77, 81, 87, 88, 89, 92—93, 97, 103, 112, 113, 114, 115, 118, 119, 123, 124, 125, 126, 127, 128, 129, 130, 133, 134.)

Hanham, two miles from Bitton, out of which it was made into an ecclesiastical parish in 1842. There is a mansion here called Hanham Court with a little ancient church adjoining.

Hawkesbury (1597), at the foot of the hill three miles south-east from Wickwar. On the top of the hill is a column—a landmark for the over Severn country—erected in 1846 in honour of Lord Edward Somerset, for 28 years (1803–31) member of Parliament for the county.

Henbury (2062), containing the tithings of Stowick, Compton, Charlton, King's Weston, is four miles from Bristol. Near the church is a ruined chapel perhaps built by William Bleys, Bishop of Worcester (1218–36), to whom the place belonged. It has been suggested that he gives his name to the modern "Blaize Castle," but the hill on which it is built was probably the site of a beacon (Blaze hill) long before.

Horsley (1079), a large village five miles from Stroud on the road to Wootton-under-Edge.

Iron Acton (1048), a village on the edge of the Bristol coalfield four miles north-west of Chipping Sodbury. Acton Court, now a farmhouse, was the residence of the family of Poyntz, one of the most ancient in the county, and still surviving elsewhere. The village gets its first name from the iron works which once existed here. (p. 74.)

Kingstanley (1877) with the ecclesiastical district of Selsey is two and a half miles from Stroud.

Kingswood (12,957), an ecclesiastical district formed out of Bitton. It is the centre of the Bristol coal trade, and must not be confused with Kingswood near Wootton-under-Edge, formerly an outlying island of Wiltshire. (p. 134.)

Lechlade (1167), a small town at the confluence of the Coln and Leach with the Thames. The church is a fine example of fifteenth century Perpendicular architecture. Half-a-mile from the town is St John's Bridge, said to be the second stone

Lechlade from the River

bridge built over the Thames, London Bridge being the first. At this point the counties of Gloucester, Berks, Wilts, and Oxford unite. At Lechlade is the eastern extremity of the Thames and Severn canal. (pp. 6, 17, 114, 118.)

Leckhampton is now incorporated in the borough of Cheltenham. There are large freestone quarries on the slope of Leckhampton Hill. (p. 29.)

Lydbrook. See Newland.

Lydney (4593), including the tithing of Aylburton, a thriving town six miles below Newnham, with an iron and tin factory and an export trade in Forest coal and iron. It has a fine Early English church with a lofty spire, and at "The Cross" is the base of a magnificent fourteenth century cross with a modern shaft. There is the base of a similar cross at Aylburton. (pp. 38, 122, 123.)

Mangotsfield, on the Midland Railway, five miles from Bristol and in the Bristol coalfield. From the station there is a branch to Bath. The population of the civil parish (which includes certain modern ecclesiastical parishes) is 9936. (pp. 28, 117, 122.)

Marshfield (1189), a large village in the south-eastern corner of the county, eight miles from Bath. (p. 6.)

Minchinhampton. A small town facing south-east, to the south of the Stroud valley. Here is a large common now used for golf. The population of the ancient, now civil, parish is 3702. (p. 122.)

Moreton-in-Marsh (1406). The last town in the county on the Fosseway, which continues its course in a north-easterly direction towards Lincoln. About two miles east on the London Road is the junction of the four counties of Gloucester, Warwick, Worcester, and Oxford, and the place is marked by a monument called the "Four Shire Stone." (pp. 5, 111, 114, 133.)

Nailsworth (3148), one of the manufacturing villages on the Bath road near Minchinhampton. It has cloth, silk, and other mills and also engineering works.

Newent (2485). The capital of the Ryeland district, nine miles from Gloucester. The nave of the church, which had been destroyed by the falling of the roof, was re-built in 1675. The chancel, Lady chapel, tower, and spire belong to the fourteenth century. (pp. 15, 36, 38, 61, 116, 122, 133.)

Newland (2203), a large parish formed in the reign of Edward I out of the borders of the Forest, but now many parts have been cut off to form ecclesiastical parishes. The village is beautifully situated in a hollow on the summit of the steep bank of the Wye. The church is a most interesting fourteenth century building. There are almshouses founded in 1615 for eight men and eight women by William Jones of Monmouth. Of the parts cut off Bream has been already mentioned; Coleford (2604), cut off in 1872, is largely inhabited by miners and is the terminus of the Monmouth and Coleford branch railway, on which Newland has a station; Clearwell, with a fourteenth century village cross, was formed in 1856; and Lea Bailey was formed out of other parishes, as was Lydbrook on the northern extremity of the Forest. (pp. 5, 15, 116.)

Newnham (1181), formerly a borough returning one member to Parliament, is a pretty town on the right bank of the Severn: the original church, which stood on the highest point of the town, commanding a beautiful prospect, was burnt down in 1881, and has been rebuilt on the same site. (pp. 20, 39, 125.)

Northleach (639), on the Cheltenham road just where it crosses the Fosse, and once one of the great wool centres of the Cotswolds, is now a forlorn and desolate town, the impetus given by the coaching days having vanished. But the splendid fifteenth century church makes amends for all. The interior resembles that of Campden, and here on the floor of the nave are the brasses of the wealthy woolstaplers, the builders of the church, such as Thomas and John Forty, and William Scots. (pp. 17, 67, 100, 101, 108, 114, 122.)

Oldland (1880), a piece of Bitton formed into an ecclesiastical parish in 1861. A school was founded here by Wesley.

Old Sodbury (763), on the high road from Gloucester to Bath, three miles from Chipping Sodbury. The Cross Hands is a famous old coaching inn.

Olveston (1406), three miles south-west from Thornbury. Within a circuit of four miles are the parish churches of Alveston, Olveston, Elberton, and Littleton-on-Severn.

Painswick (2638), a clean town with many fine houses on the high ground between Stroud and Cheltenham. There is a large fifteenth century church with a tall spire. The churchyard is remarkable for its beautifully carved tombstones, and its rows of clipped yews. On the summit of Painswick hill, a mile and a half to the north of the town, at an altitude of 900 feet, is Kimsbury camp, from which there is a magnificent prospect from Radnor Forest to the mouths of the Avon and the Wye, and from the Clee Hills to the Berkshire White Horse. (pp. 100, 101.)

Prestbury (1806), a pleasant village two miles north-east of Cheltenham.

Pucklechurch (1298), a village three miles east of Mangotsfield.

Rodborough (3721), a village on the opposite side of the valley to Stroud. Rodborough Fort is a modern building on the extremity of the spur, and conspicuous from the other side of the Severn.

Ruardean (1273), a village at the top of a hill with a tall church tower looking across the Wye into Herefordshire.

St Briavels (1128), loftily situated on the summit of the steep bank of the Wye opposite Trelleck. The castle, which is still inhabited, was the residence of the Constable of the Forest. The interesting cruciform church has lost its central tower, and the modern tower is very inferior. (pp. 38, 104, 124.)

Slad, The, a piece of Painswick formed into an ecclesiastical parish in 1844.

Stonehouse (2304), a clean well-built village three miles from Stroud. The inhabitants are engaged in the various branches of the clothing trade. (p. 122.)

Stow-on-the-Wold (1204). The highest town on the Cotswolds (nearly 800 feet): the church tower is visible for ten miles round. Six roads converge here and all except the one from Broadway have to climb a steep hill to reach the town. To the outside world the town is best known for its fairs, which take place on 12 May and 24 October. (pp. 82, 108.)

Stroud (8561). The chief seat of the clothing trade. It originally formed part of the parish of Bisley, but was separated from it in the reign of Edward II. The town lies on the northern slope of the Frome or Stroudwater valley, just where it opens out into a wide basin, and is visible from Tidenham on the other side of the Severn. The woollen manufactures have existed here ever since the English first began to weave woollen cloth. The West of England broadcloth is famous throughout the British Isles. (pp. 32 n., 65, 67, 68, 69, 125.)

Tetbury (1593). In spite of a branch railway from Kemble, this little town has declined rapidly in recent years. It has a church rebuilt in 1781 and a market-place where corn is sold. Two miles distant are the ruins of Beverston Castle, a fourteenth century edifice. The Bristol Avon rises close to the town. (pp. 22, 60, 114.)

Tewkesbury (4704). A flourishing town at the junction of the Avon with the Severn. There were formerly nail factories and tanneries here, but the industries are now confined to flour-mills and a small shirt factory. There are many fine old half-timbered houses, of which the Bell Inn, with its splendid bowling green, is the most remarkable. But the town is best known as the site of the victory of Edward IV over Queen Margaret and the Lancastrians in 1471, and for its magnificent

Tewkesbury Abbey

parish church, which before the Dissolution was the church of the great Benedictine monastery of St Mary. The nave with its imposing Norman arcade may be compared with that of Gloucester. The central tower is also Norman; the choir is early fourteenth century. The church is remarkable for its many beautiful monuments dating from the fourteenth to the sixteenth century.' Tewkesbury is one of the five Gloucestershire boroughs. (pp. 8, 18, 20, 30, 77—79, 88, 95—97, 100, 122, 124, 125.)

Thornbury (2646). A declining town eleven miles north of Bristol with a branch railway from Yate Junction. The castle, an interesting example of early Tudor architecture, was begun in 1511 by Edward Stafford third Duke of Buckingham who was beheaded in 1521, but it was never finished. The fifteenth century church has a fine lofty tower—a landmark from the other side of the water. (pp. 117, 125.)

Tidenham (1710), a large parish forming the extremity of the Forest peninsula. The extreme point is Beachley, formed into an ecclesiastical parish in 1850. At the north side of the triangle is Tidenham Chace, which reaches the height of 778 feet. The hamlet of Stroat is so called from the Roman road (street) on which it lies. The church is a landmark from the other side of the water. On the western side of the parish is Tutshill, with the remains of an ancient watch-tower from which the approach of strange vessels in the Wye could be signalled to Chepstow Castle. In the Wye cliffs near the little peninsula of Lancaut are large stone quarries—a sad disfigurement of the quiet landscape. This parish, together with the adjoining parish of Woolaston, were part of the Lordship Marcher of Striguil (Chepstow) till the formation of the county of Monmouth in 1534, when they were added to Gloucestershire. They had, however, always belonged to the diocese of Hereford. (pp. 13, 22, 30, 36, 37, 40, 41, 61, 124.)

Upton St Leonard's (1124), a parish three miles from Gloucester, on the Painswick road.

Westbury-on-Severn (1791), a village on the right bank of the Severn nine miles from Gloucester.   The church is sixteenth century, but the detached tower is earlier.   Here is Framilode passage across the river.   (p. 30)

Yate Church

Westerleigh (1128), in the Bristol coalfield; parts of it have been taken to form other ecclesiastical parishes, the population of the present ecclesiastical parish being only 429.

Whiteshill (1342), formed into an ecclesiastical parish in 1844 out of Standish and Stroud. In 1894 it was constituted a civil parish with a larger population.

Wickwar (860). A small town, formerly a borough, celebrated for its breweries, 15 miles NNE. from Bristol.

Winchcombe (2930). A small town at the foot of the Cotswolds, seven miles north-east of Cheltenham. It was formerly of great importance owing to its wealthy Benedictine abbey, not a vestige of which beyond the foundations is left. Sudeley Castle, once the home of the Chandos family, is on the opposite side of the Isbourne. (pp. 19, 88, 116).

Winterbourne (3191) is a large parish in the Bristol coalfield some four miles north of Bristol. In 1836 it was divided into two ecclesiastical parishes.

Wootton-under-Edge (3021), a small declining town with a station on the Midland Railway. It lies at the foot of the hills, on the summit of which is Lord Edward Somerset's monument. The grammar school, which still flourishes, is the oldest in England, having been founded in 1382. (p. 32 n.)

Yate (1309). A village ten miles north-east of Bristol with a station on the Midland Railway and a branch line to Thornbury. (117.)

Fig. 1. Area of Gloucestershire (805,842 acres) compared
with that of England and Wales

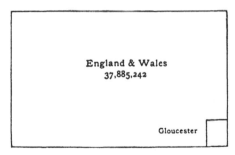

Fig. 2. Population of Gloucestershire (757,668) compared
with that of England and Wales in 1921

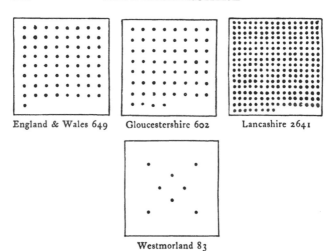

England & Wales 649   Gloucestershire 602   Lancashire 2641

Westmorland 83

Fig. 3. Comparative density of Population to the
square mile in 1921

(*Each dot represents 10 persons*)

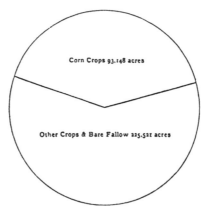

Corn Crops 93,148 acres

Other Crops & Bare Fallow 225,521 acres

Fig. 4. Proportionate area under Corn Crops compared with
that of other cultivated land in Gloucestershire in 1923

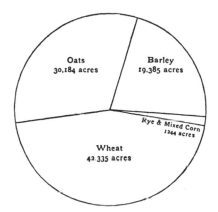

Fig. 5. Proportionate areas of chief Cereals in
Gloucestershire in 1923

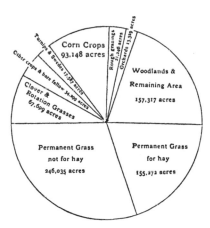

Fig. 6. Proportionate areas of land in
Gloucestershire in 1923

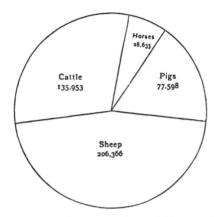

Fig. 7. Proportionate numbers of Live Stock in
Gloucestershire in 1923

Milton Keynes UK
Ingram Content Group UK Ltd.
UKHW041520181024
449640UK00009B/85

9 781107 697393